Alan McSeveny Rachel Adams Diane McSeveny-Foster

Pearson Australia
(a division of Pearson Australia Group Pty Ltd)
459–471 Church St, Level 1, Building B, Richmond, Victoria, 3121
PO Box 23360, Melbourne, Victoria 8012
www.pearson.com.au

First published 2023 by Pearson Australia
2026 2026 2025 2024
10 9 8 7 6 5 4 3 2 1

Publishers: Sophie Matta and Rachel Elliot
Project Manager: Michelle Thomas
Production Editor: Laura Rentsch
Editor: Katie Miller
Designer: Anne Donald
Proofreader: Laura Rentsch
Rights & Permissions Editor: Alice McBroom
Cover Design: Jennifer Johnston
Cover Art: Michael Barter
Illustrator: Michael Barter
Desktop Operator: Jit-Pin Chong
Printed in Malaysia by Vivar

National Library of Australia Cataloguing-in-Publication entry

A catalogue record for this book is available from the National Library of Australia

ISBN 978 0 6557 0901 5

Pearson Australia Group Pty Ltd ABN 40 004 245 943

Acknowledgement of Country
Pearson respects and honours Aboriginal and Torres Strait Islander Elders past, present and future. We acknowledge the stories, traditions and living cultures of the Traditional Custodians of the lands on which our company is located and where we conduct our business. Pearson is committed to honouring Australian Aboriginal and Torres Strait Islander peoples' unique cultural and spiritual relationships to the land, waters and seas and their rich contribution to society.

Aboriginal and Torres Strait Islander peoples are advised that this text may contain images, voices and names of deceased persons.

What is Australian Signpost Maths NSW?

Australian Signpost Maths NSW is a mathematics program providing direction and support for teaching and learning. The series covers the content and skills presented in the NSW Mathematics Syllabus K–6, 2022.

A Student Book and an online Teacher Resource are provided for Kindergarten (Early Stage 1).

For Years 1 to 6 (Stages 1–3), a Student Book, an online Teacher Resource and a Mentals Book are provided for each year level. The online Teacher Resources provide a wealth of support for teachers.

The content has been carefully sequenced within each year level and across the K–6 series to take into account students' expected mathematical development. However, from the rich and varied material provided, teachers can develop individual learning programs to meet the needs of each student.

The Student Books are designed to support explicit teaching methods. Many group activities are provided in Activity, Investigation and Fun spots within the Student Books and the online Teacher Resource.

To maximise the benefits of the program, the Student Book, the online Teacher Resource and the Mentals Book should be used together.

Student Books

Mentals Books

Teacher Resource

Structure of Australian Signpost Maths NSW

In the K–2 books, the worksheet pages covering all three strands are presented in a recommended order. Each unit of 4 pages usually begins with Number and algebra. The Contents cross-reference allows teachers to quickly find the pages where each concept has been covered.

Within the program, explicit teaching, working mathematically skills, language development and identification and treatment of weaknesses are given high priority.

Identification and addressing areas of need

Five progress tests are designed to identify each student's areas of need, and the follow-up program after each of the tests is designed to address these needs. Beside each test question, a reference to the relevant worksheet page is given. A remediation record page is used to track the student's progress.

These testing resources can be found in the online Teacher Resource.

Parallel progress retests are provided for further testing after remediation has taken place. See pages 128 and 129 for more information.

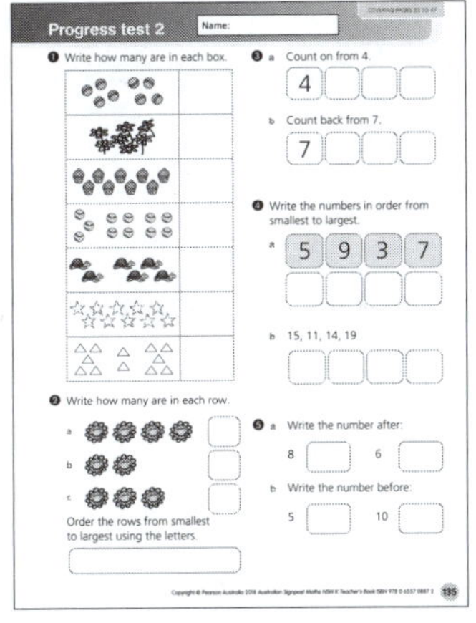
Progress test 2 Name:

1 Write how many are in each box.

2 Write how many are in each row.

a

b

c

Order the rows from smallest to largest using the letters.

3 a Count on from 4.

4

b Count back from 7.

7

4 Write the numbers in order from smallest to largest.

a 5 9 3 7

b 15, 11, 14, 19

5 a Write the number after:

8 6

b Write the number before:

5 10

135

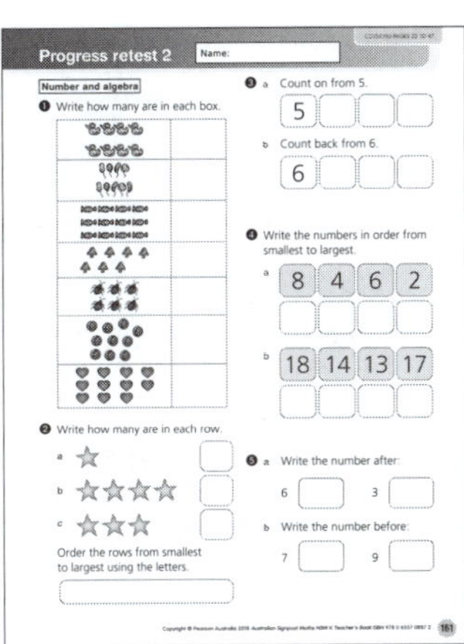
Progress retest 2 Name:

Number and algebra

1 Write how many are in each box.

2 Write how many are in each row.

a

b

c

Order the rows from smallest to largest using the letters.

3 a Count on from 5.

5

b Count back from 6.

6

4 Write the numbers in order from smallest to largest.

a 8 4 6 2

b 18 14 13 17

5 a Write the number after:

6 3

b Write the number before:

7 9

161

Special features of Australian Signpost Maths NSW

- **The traffic light icons**
  These are found on the top right of each worksheet page in the Student Books. They allow students to assess their own progress and give feedback to the teacher.
  - **Green:** I found this work easy.
  - **Orange:** I found some work on the page difficult.
  - **Red:** I don't understand the work on this page.

- **Dictionary**
  Terms used in the Student Book and terms that should be understood at this level are recorded here to provide a reference for students and teachers. This is found on pages xii–xiv of this book and in the online Teacher Resource.

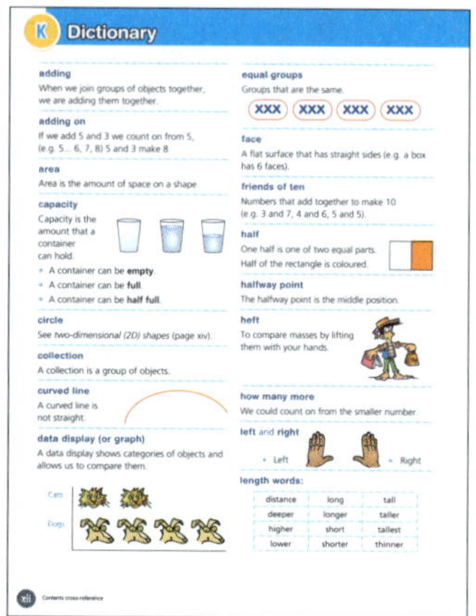

Dictionary

adding
When we join groups of objects together, we are adding them together.

adding on
If we add 5 and 3 we count on from 5, (e.g. 5... 6, 7, 8) 5 and 3 make 8.

area
Area is the amount of space on a shape.

capacity
Capacity is the amount that a container can hold.
- A container can be **empty**.
- A container can be **full**.
- A container can be **half full**.

circle
See *two-dimensional (2D) shapes* (page xiv).

collection
A collection is a group of objects.

curved line
A curved line is not straight.

data display (or graph)
A data display shows categories of objects and allows us to compare them.

Cats
Dogs

equal groups
Groups that are the same.
XXX XXX XXX XXX

face
A flat surface that has straight sides (e.g. a box has 6 faces).

friends of ten
Numbers that add together to make 10 (e.g. 3 and 7, 4 and 6, 5 and 5).

half
One half is one of two equal parts.
Half of the rectangle is coloured.

halfway point
The halfway point is the middle position.

heft
To compare masses by lifting them with your hands.

how many more
We could count on from the smaller number.

left and **right**
- Left
- Right

length words:

distance	long	tall
deeper	longer	taller
higher	short	tallest
lower	shorter	thinner

xii Contents cross-reference

- **ID cards (Years 1 to 6)**
  These cards review the language of Mathematics by asking students to identify common terms, shapes and symbols. They are designed to be reused and are found in the online Teacher Resource and in the front of the Mentals Books.

- **Progress tests**
  These allow the teacher to identify each student's strengths and needs. The cross-references beside each question direct teachers and students to the pages where that work is introduced. Tables are provided to record the follow-up that takes place and parallel tests are provided for retesting. These tests can be found in the online Teacher Resource.

- **Year K Consolidation Booklet**
  This 30 page booklet is found in the online Teacher Resource. It is designed to reinforce work completed in class and provides practice of important skills and addition and subtraction facts. The booklet can be used when there is limited supervision or when a student finishes classwork early.

Signpost K
Consolidation booklet

Student's name:
Class:

- **Answers**
  These are supplied in the online Teacher Resource.

- **Blackline Masters (BLM)**
  References are made to the Blackline Masters in the teaching suggestions provided for each student work page.

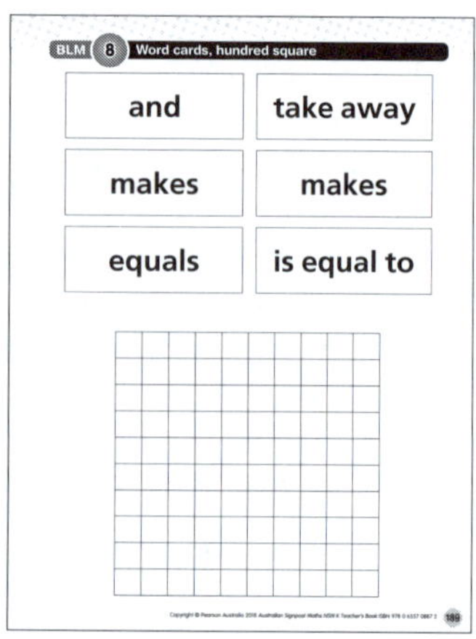

BLM 8 Word cards, hundred square

and	take away
makes	makes
equals	is equal to

- **Differentiation**
  Each student work page has a Teacher Resource page to support it. Cross-references direct the teacher to pages where the concept is introduced and developed. These references may be from the Student Book for the previous year, current year or the next year.

  The Teacher Resource support pages provide additional learning activities for students who need remediation or extension activities. The Blackline Master Worksheets provide activities to support students of various learning abilities.

- **Cartoons**
  Cartoons are used to motivate and instruct.

Australian Signpost Maths NSW icons

Signpost icons are used throughout the book as cues to the essential nature of exercises and activities, and as a guide to ways of engaging with them. These icons often indicate alternative or more concrete approaches to dealing with concepts.

Cutting and pasting.

Drawing, matching and colouring.

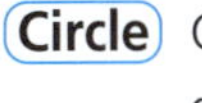

Circling the best or correct answer.

This icon highlights **important rules and concepts** occurring throughout the book. It often appears with worked examples.

Investigations allow students to **explore and discover** maths concepts.

Activities provide **applications and enrichment**. These activities usually involve the use of concrete materials and partner or group work.

This icon indicates the use of computers, calculators or other **information and communications technology**.

These enjoyable activities are used to **motivate and involve** students in mathematical pursuits. They usually involve games and puzzles.

Structure of New South Wales Mathematics K–6

The NSW Mathematics Syllabus content is presented in three strands.

1 Number and algebra
2 Measurement and space,
3 Statistics and probability.

Working mathematically pervades each of these strands.

The Mathematics Syllabus can be found at:
https://curriculum.nsw.edu.au/learning-areas/mathematics/mathematics-k-10

Textbook Structure
Within the Contents for Kindergarten, we show related pages using these categories:

Number and algebra	**Measurement and space**	**Statistics and probability**
Numbers	2D shapes / 3D objects	Data displays
Addition / subtraction	Length / area / mass	
Sharing / grouping	Capacity / volume	
Patterns	Time / duration	
	Position	

Contents and syllabus overview

KEY

Colour	Strand
Blue	Number and algebra
Teal	Measurement and space
Orange	Statistics and probability

Page	Unit	Title	Strand	Number/algebra	Measurement/space	Statistics/probability	Content area	Numbers	Addition/subtraction	Sharing/grouping	Patterns	2D Shapes/3D objects	Length/area/mass	Capacity/volume	Time/duration	Position	Data displays
1	Thinking Skills		Working mathematically pervades each of the strands.														
2	1A	Zero		■				●									
3	1B	The number one		■				●									
4	1C	The number two		■				●									
5	1D	Long, short and tall			■								●				
6	2A	The number three		■				●									
7	2B	The number four		■				●									
8	2C	The number five		■				●									
9	2D	Data				■											●
10	3A	Numbers to five		■				●									
11	3B	Counting to five		■				●									
12	3C	The number six		■				●									
13	3D	Curved and straight			■							●					
14	4A	The number seven		■				●									
15	4B	Dot patterns		■				●									
16	4C	Circles			■							●					
17	4D	Heavy or light			■								●				
18	5A	Same and not the same		■				●									
19	5B	Same and not the same		■				●									
20	5C	Squares			■							●					
21	5D	Full, empty and half full			■									●			
		Progress Test 1: Administer test (Teacher Resource, pages 131–133) then address weaknesses.															
22	6A	The number eight		■				●									
23	6B	Comparing groups		■				●									
24	6C	Triangles			■							●					
25	6D	Comparison of mass			■								●				
26	7A	The number nine		■				●									
27	7B	The number ten		■				●									
28	7C	Rectangles			■							●					
29	7D	Cutting shapes			■							●					

KEY

- Number and algebra
- Measurement and space
- Statistics and probability

Page	Unit	Title	Strand	Number/algebra	Measurement/Space	Statistics/probability	Content area	Numbers	Addition/subtraction	Sharing/grouping	Patterns	2D Shapes/3D objects	Length/area/mass	Capacity/volume	Time/duration	Position	Data displays
30	8A	Numbers to ten		■				●									
31	8B	Numbers to ten		■				●									
32	8C	Position			■											●	
33	8D	Language of location			■											●	
34	9A	Numbers to 10		■				●									
35	9B	Numbers 11 and 12		■				●									
36	9C	Longer and shorter			■								●				
37	9D	Daytime and night-time			■										●		
38	10A	Adding two groups		■					●								
39	10B	Adding two groups		■					●								
40	10C	Shape pictures			■							●					
41	10D	Morning and afternoon			■										●		
42	11A	Numbers 13 to 20		■				●									
43	11B	Numbers 11 to 20		■				●									
44	11C	Sorting objects			■							●					
45	11D	3D objects			■							●					
46	12A	Addition		■					●								
47	12B	Adding two groups		■					●								

Progress Test 2: Administer test (Teacher Resource, pages 135–138) then address weaknesses.

Page	Unit	Title	Strand	Number/algebra	Measurement/Space	Statistics/probability	Content area	Numbers	Addition/subtraction	Sharing/grouping	Patterns	2D Shapes/3D objects	Length/area/mass	Capacity/volume	Time/duration	Position	Data displays
48	12C	Rolling, sliding and stacking			■							●					
49	12D	Using data displays				■											●
50	13A	Using five to form numbers		■					●								
51	13B	Adding two groups		■					●								
52	13C	Using data displays				■											●
53	13D	Sequencing events in a day			■									●			
54	14A	Adding two groups		■					●								
55	14B	Adding groups		■					●								
56	14C	Comparing distances			■								●				
57	14D	Indirect comparison			■								●				
58	15A	Ordering collections		■							●						
59	15B	Halves			■								●				
60	15C	Clocks			■										●		
61	15D	Telling the time			■										●		
62	16A	Ordinal numbers			■											●	
63	16B	Patterns using sounds and actions		■							●						
64	16C	Using o'clock			■										●		
65	16D	Digital time			■										●		

KEY

■	Number and algebra
■	Measurement and space
■	Statistics and probability

Page	Unit	Title	Strand	Number/algebra	Measurement/Space	Statistics/probability	Content area	Numbers	Addition/subtraction	Sharing/grouping	Patterns	2D Shapes/3D objects	Length/area/mass	Capacity/volume	Time/duration	Position	Data displays
66	17A	Adding groups		■					●								
67	17B	Adding groups		■					●								
68	17C	Ball-shaped objects			■							●					
69	17D	Sorting objects			■							●					
70	18A	Taking objects away		■					●								
71	18B	Taking away		■					●								
72	18C	Area			■								●				
73	18D	Days of the week			■										●		
74	19A	Taking away		■					●								
75	19B	Taking away		■					●								
76	19C	Comparing two lengths			■								●				
77	19D	Patterns		■							●						
78	20A	Separating a number into parts		■					●								
79	20B	Separating a number into parts		■					●								
80	20C	Shapes			■							●					
81	20D	Data displays				■											●
82	21A	Counting to 20		■				●									
83	21B	Comparing collections		■				●									

Progress Test 3: Administer test (Teacher Resource, pages 140–143) then address weaknesses.

Page	Unit	Title	Strand	Number/algebra	Measurement/Space	Statistics/probability	Content area	Numbers	Addition/subtraction	Sharing/grouping	Patterns	2D Shapes/3D objects	Length/area/mass	Capacity/volume	Time/duration	Position	Data displays
84	21C	Cone-shaped objects			■							●					
85	21D	Left and right			■											●	
86	22A	Groups of equal size		■						●							
87	22B	Matching equal groups		■						●							
88	22C	Box-shaped objects			■							●					
89	22D	Length			■								●				
90	23A	Rearranging groups		■				●									
91	23B	Using groups to share		■						●							
92	23C	Shapes			■							●					
93	23D	Data displays				■											●
94	24A	Counting to 30		■				●									
95	24B	Adding on and counting back		■					●								
96	24C	Can-shaped objects			■							●					
97	24D	Lighter or heavier			■								●				
98	25A	Taking away		■					●								
99	25B	Taking away		■					●								
100	25C	Giving and following instructions			■											●	
101	25D	Sequencing events			■										●		

KEY

Colour	Strand
Blue	Number and algebra
Green	Measurement and space
Orange	Statistics and probability

Page	Unit	Title	Strand	Number/algebra	Measurement/Space	Statistics/probability	Content area	Numbers	Addition/subtraction	Sharing/grouping	Patterns	2D Shapes/3D objects	Length/area/mass	Capacity/volume	Time/duration	Position	Data displays
102	26A	How many more?		■					●								
103	26B	How many more?		■					●								
104	26C	Stacking and packing			■							●					
105	26D	Volume			■									●			
106	27A	How many more?		■					●								
107	27B	Sharing		■						●							
108	27C	Classifying 2D shapes			■							●					
Progress Test 4: Administer test (Teacher Resource, pages 145–148) then address weaknesses.																	
109	27D	Days of the week			■										●		
110	28A	Sharing		■						●							
111	28B	Sharing among 3 or more		■						●							
112	28C	Comparison of areas			■								●				
113	28D	Comparing areas			■								●				
114	29A	Sharing in other ways		■						●							
115	29B	Comparing lengths			■								●				
116	29C	Comparing internal volumes			■									●			
117	29D	Comparing internal volumes			■									●			
118	30A	The halfway point			■								●				
119	30B	Halfway			■								●				
120	30C	Comparing internal volumes			■									●			
121	30D	Volume using blocks			■									●			
122	31A	2D shapes			■							●					
Progress Test 5: Administer test (Teacher Resource, pages 150–143) then address weaknesses.																	
123	31B	Using data displays				■											●
124	31C	Describing objects in our world			■							●					
125	31D	Recording the weather				■											●
126	32A	Location			■											●	
127	32B	Pattern blocks			■							●					
128	Identifying and addressing areas of need																
130	**1** Number charts	**2** Ten frames	**3** Place value cards														
133	**4** Adding two groups	**5** Finding the difference	**6** Sharing														
136	**7** Number bond houses	**8** Number bonds (addition)	**9** Using number tracks														

Contents cross-reference

Number and algebra

Measurement and space

1	Geometric measure	Pages
	Position: Describing position	5, 32, 33, 85, 89, 126
	Giving and following directions	56, 100
	Ordinal numbers	62, 101
	Left and right	85, 89, 100, 126
	Length: Describing and comparing lengths	5, 36, 56, 57, 76, 89, 115
	About half a length, more than, less than half a length, halfway point	59, 118, 119
2	**Two-dimensional spatial structure**	**Pages**
	2D shapes: Sorting, describing and making shapes	16, 20, 22, 24, 28, 29, 40, 80, 92, 108, 122, 127,
	Circles, squares, triangles and rectangles	16, 20, 22, 24, 28, 29, 40, 80, 92, 108, 122, 127
	Straight lines and curved lines	13, 16, 29, 108, 122
	Area: Describing and comparing areas	72, 112, 113
3	**Three-dimensional spatial structure**	**Pages**
	3D objects: Describing and sorting 3D objects	44, 45, 48, 68, 69, 84, 88, 96, 124
	Stacking 3D objects	44, 45, 48, 68, 69, 84, 88, 96, 104, 124
	Volume: Describing and comparing volumes	104, 105, 121
	Full, empty and about half full	21, 116
	Comparing internal volumes	104, 105, 116, 117, 120
	Stacking, packing and building to compare volumes	104, 105, 121
4	**Non-spatial measure**	**Pages**
	Mass: Describing and comparing the weight of objects	17, 25, 97
	Time: Describing and comparing times, sequencing events	37, 41, 53, 101
	Days of the week	73, 101, 109
	Telling time on the hour using analog and digital clocks	37, 41, 53, 60, 61, 64, 65

Statistics and probability

1	Data	Pages
	Collecting information	9, 49, 81, 93, 123, 125
	Using data displays	23, 49, 52, 81, 93, 123, 125

Dictionary

K

adding

When we join groups of objects together, we are adding them together.

adding on

If we add 5 and 3 we count on from 5, (e.g. 5... 6, 7, 8) 5 and 3 make 8.

area

Area is the amount of space on a shape.

capacity

Capacity is the amount that a container can hold.

- A container can be **empty**.
- A container can be **full**.
- A container can be **half full**.

circle

See *two-dimensional (2D) shapes* (page xiv).

collection

A collection is a group of objects.

curved line

A curved line is not straight.

data display (or graph)

A data display shows categories of objects and allows us to compare them.

equal groups

Groups that are the same.

face

A flat surface that has straight sides (e.g. a box has 6 faces).

friends of ten

Numbers that add together to make 10 (e.g. 3 and 7, 4 and 6, 5 and 5).

half

One half is one of two equal parts.

Half of the rectangle is coloured.

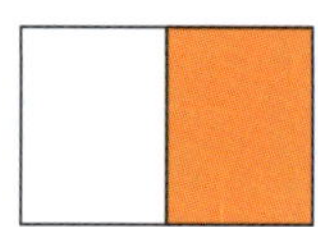

halfway point

The halfway point is the middle position.

heft

To compare masses by lifting them with your hands.

how many more

We could count on from the smaller number.

left and **right**

- Left

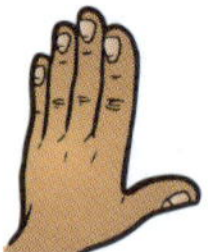
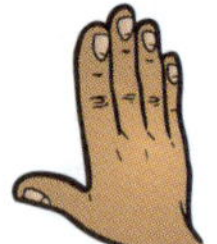

- Right

length words:

distance	long	tall
deeper	longer	taller
higher	short	tallest
lower	shorter	thinner

mass words:

heavy	light	weigh
heavier	lighter	weight

number bonds

Pairs of numbers that add to make a specific number (e.g. 4 and 0, 1 and 3, and 2 and 2 all make 4).

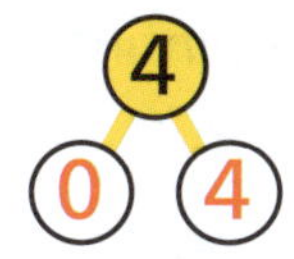

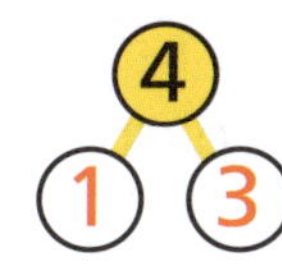

number sentence

Uses numerals and symbols (e.g. 4 and 6 makes 10. This can be written as 4 + 6 = 10).

numeral

A numeral is a written number symbol.

7, 18, 92, 120

ordering

We can order objects or numbers by placing them from smallest to largest or from largest to smallest.

We can use words like first, next, last, before, after, 1st, 2nd, 3rd.

ordinal numbers

Ordinal numbers describe the order or position of something.

1st, 2nd, 3rd, 4th, 5th

pattern

A pattern is a group of numbers, objects, shapes, colours, sounds or actions that are repeated over and over again.

rectangle

See *two-dimensional (2D) shapes* (page xiv).

row

A row is a line of objects going across. Here are 2 rows of 5.

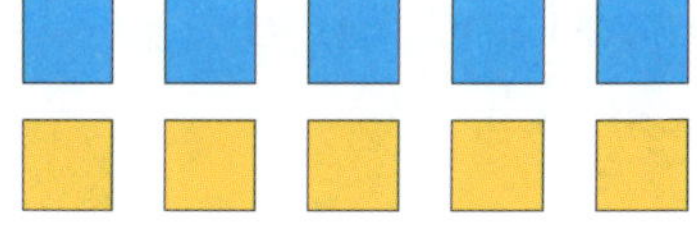

sharing

When sharing, we make sure that each share is the same size.

If 2 people could share these 6 balls, each person would get 3 balls.

If two groups are not the same, we can make fair shares by moving items from the larger group to the smaller group.

sorting, classifying

When sorting or classifying, we put similar objects in each group.

Dogs

Cats

square

See *two-dimensional (2D) shapes* (page xiv).

straight line

A straight line has no bends or curves.

take away (subtraction)

When we remove objects from a group, we call this 'take away'.

three-dimensional (3D) objects

Solid objects are three-dimensional.
They have length, width and height.

Ball-shaped objects (spheres)

are curved and round.
They can roll.

have 6 square faces.
They can slide and stack.

Can-shaped objects (cylinders)

have 2 surfaces that are circles and 1 curved surface. They can roll, slide and stack.

Cone-shaped objects (cones)

have 1 surface that is a circle and 1 curved surface. They can roll and slide.

time words

morning	daytime
afternoon	night-time

Days			
Sunday	Monday	Tuesday	Wednesday
Thursday	Friday	Saturday	

Clocks

o'clock:
When the long hand is pointing to 12, the time is an o'clock time.

triangle

See *two-dimensional (2D) shapes* (page xiv).

two-dimensional (2D) shapes

Flat shapes are two-dimensional.
They have length and width.

Circle
1 curved side

Triangle
3 straight sides

Square
4 equal straight sides

Rectangle
2 equal long sides and 2 equal short sides, like a stretched square. All sides are straight.

volume

Volume is the amount of space an object takes up.

The mice and the parrot

1. What are the mice wearing?
2. How many hats are in this picture?
3. How many balloons are in this picture?
4. What else can you count in the picture?
5. What things are round? Colour the round things.
6. What is the parrot doing?
7. What could happen when the mouse sticks the needle into the balloon?
8. Why is the mouse with balloons worried?
9. How many legs can you see in this picture?
10. Which of these questions do you like best? Why do you like it?

 AUSTRALIAN SIGNPOST MATHS NSW K • ISBN 9780655709015

1A Zero

1 Circle the containers that hold zero things.  Trace the numerals and the word "zero".

2 These stalls were at the show. Look at the picture and then answer the questions.

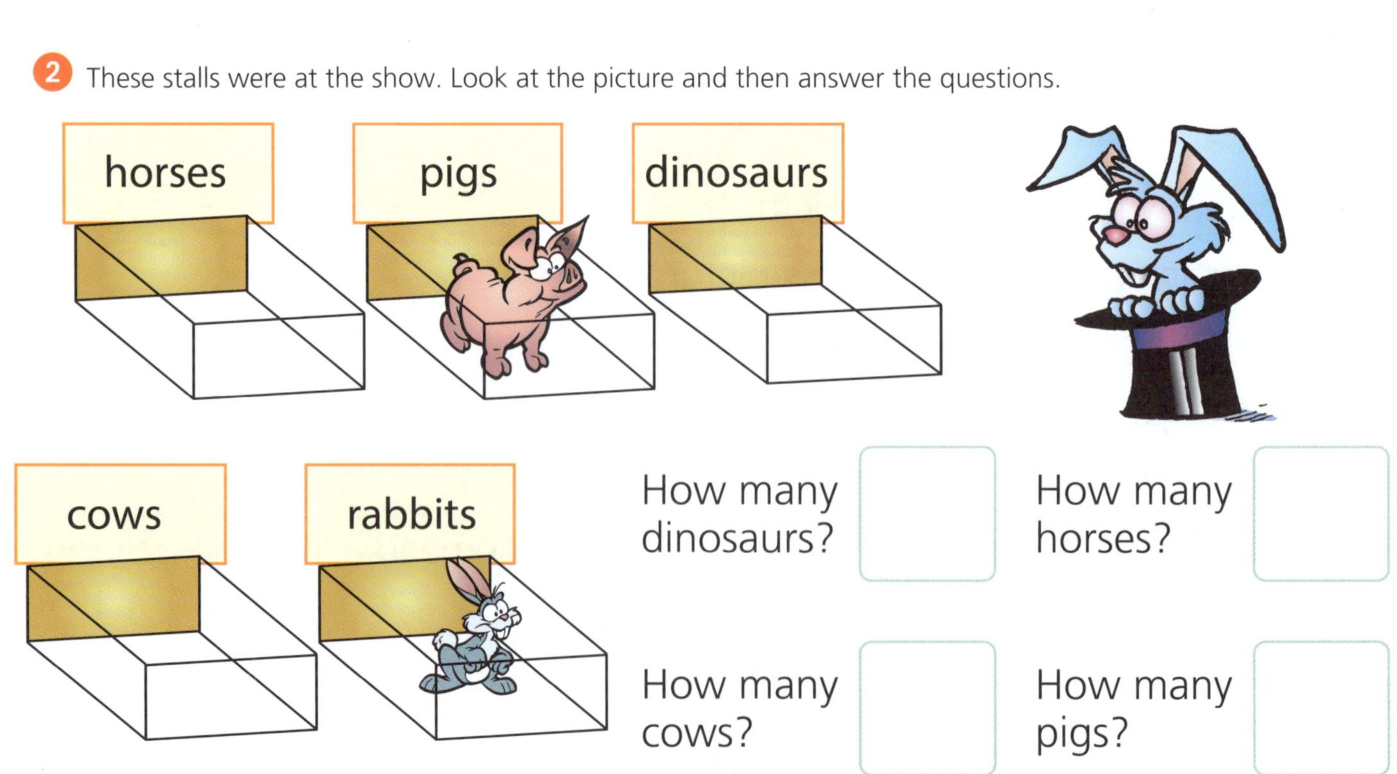

How many dinosaurs?

How many horses?

How many cows?

How many pigs?

1B The number one

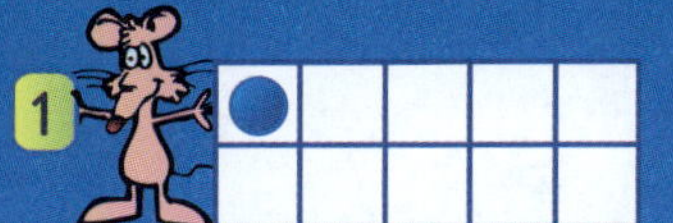

1 **Circle** the groups with one object. Trace the numerals and the word "one".

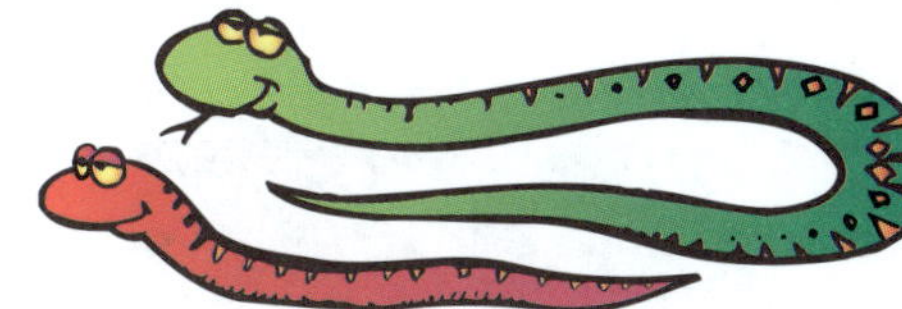

2 Draw one fish.

How many fish did you draw?

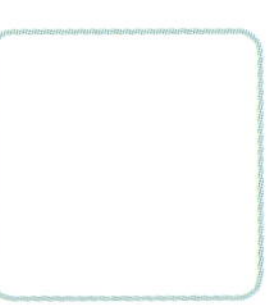

FUN SPOT

Colour the parts that are shown only once.

© PEARSON AUSTRALIA 2023 • *AUSTRALIAN SIGNPOST MATHS NSW K* • ISBN 9780655709015

1C The number two

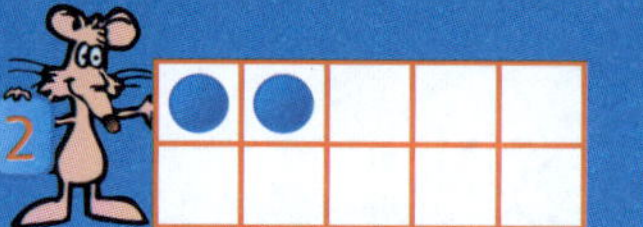

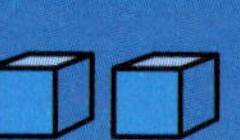

1 **Circle** the groups of two. Trace the numerals and the word "two".

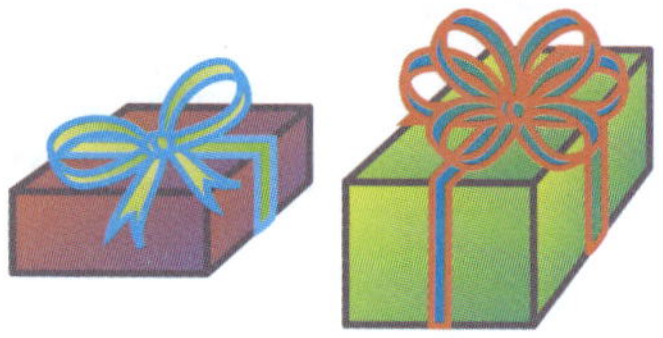

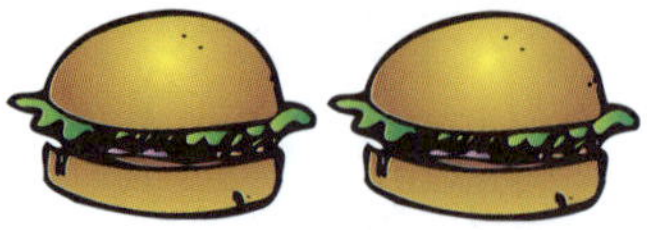

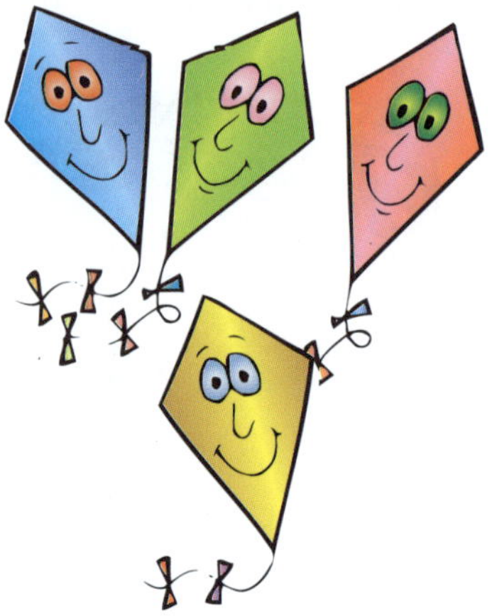

2 two

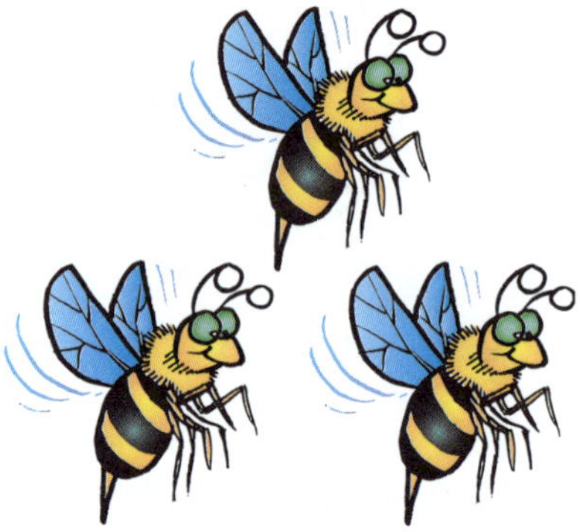

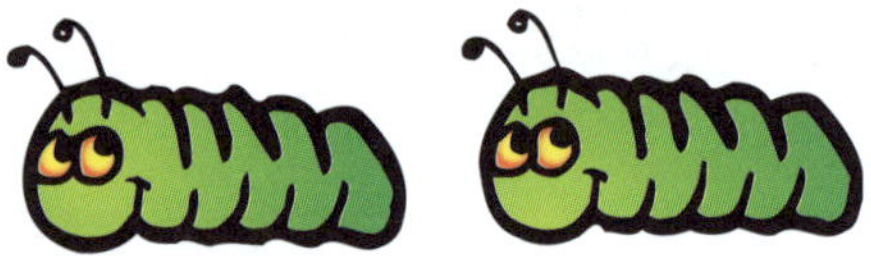

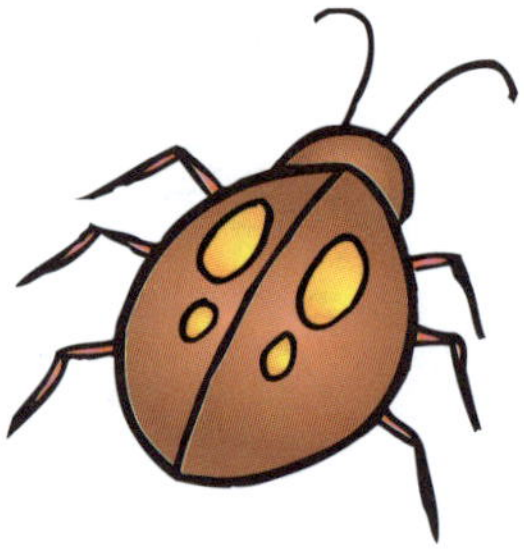

2 2 2 2 2 2

Discuss which groups above have the same number.

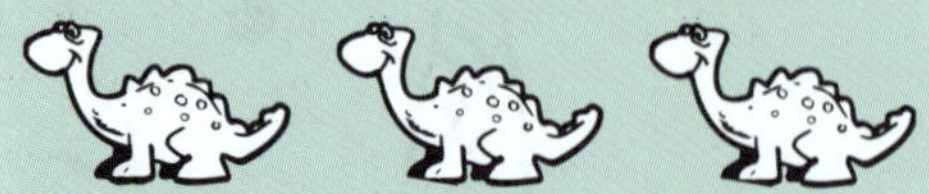

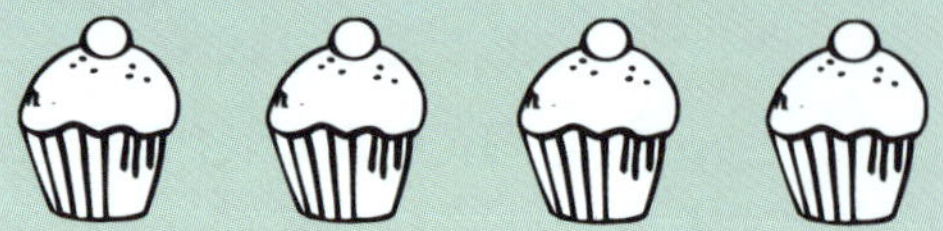

2 Colour two in each row. Trace the numbers.

Draw two balloons.

Tell a story about the balloons.

1D Long, short and tall

1 Draw lines to match each word to a picture.

short | long | tall

2 Draw a tall tree. | Draw one long arrow. | Draw a long scarf.

Draw two short trees. | Draw two short arrows. | Draw two short scarves.

3 Colour the long snakes. Draw two more short snakes.

 • *AUSTRALIAN SIGNPOST MATHS NSW K* • ISBN 9780655709015

2A The number three

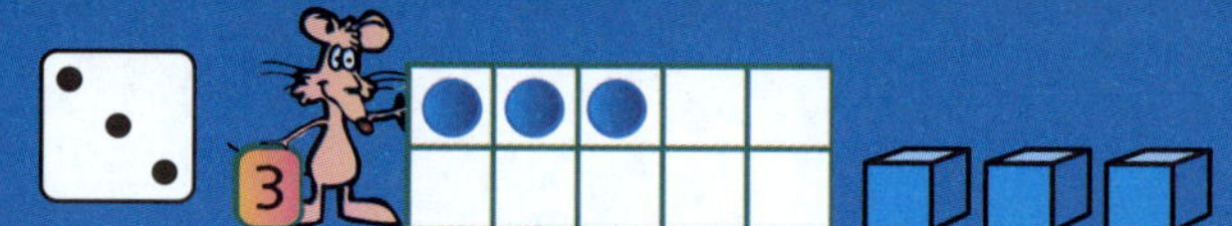
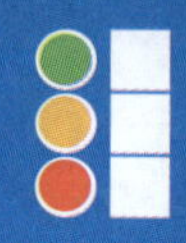

1 Colour groups that show three. Trace the numerals and the word "three".

2 Draw three spots on each ladybird.

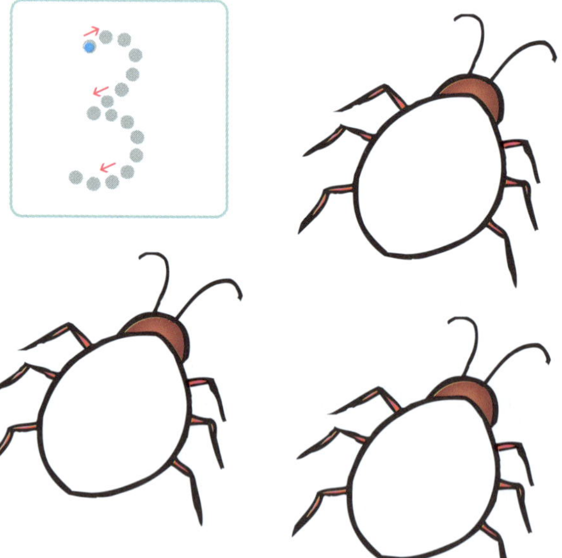

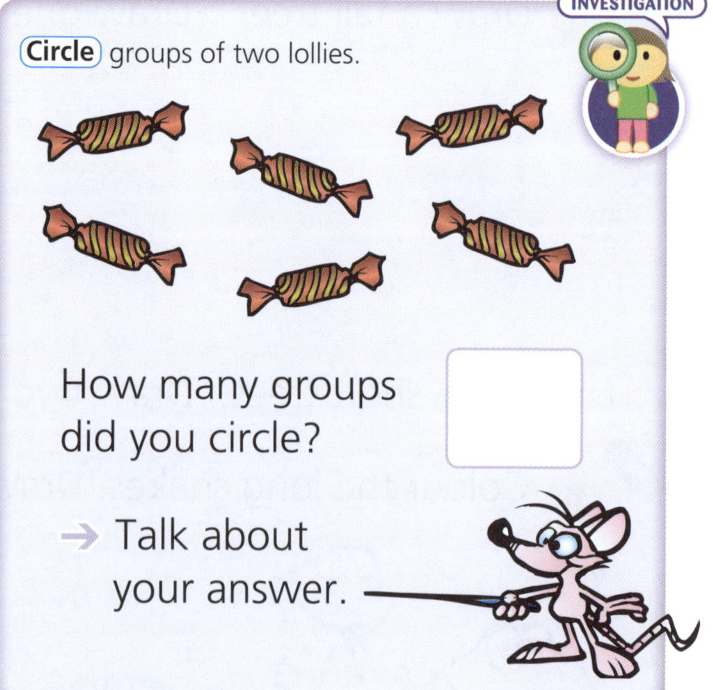

3 Write the numeral after or before.

 • *AUSTRALIAN SIGNPOST MATHS NSW K* • ISBN 9780655709015

2B The number four

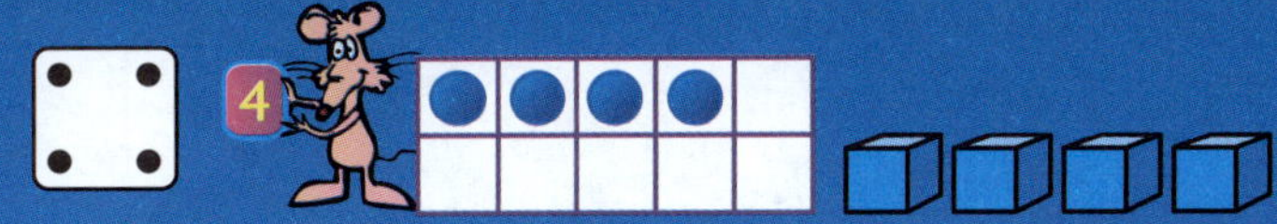

1 Circle groups that show four.

Trace the numerals and the word "four".

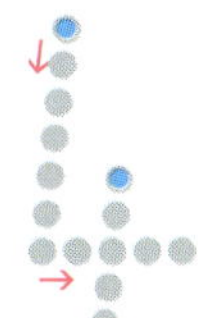

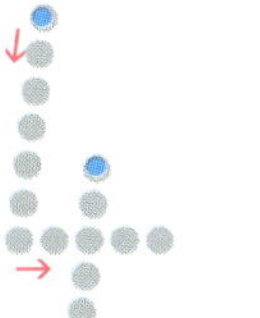

2 Draw objects to match each numeral.

INVESTIGATION

Make different dot patterns for four.

→ Talk about your answer.

3 Write the numbers 0, 1, 2, 3 and 4 from smallest to largest.

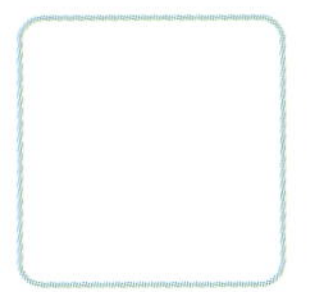

Number and algebra

2C The number five

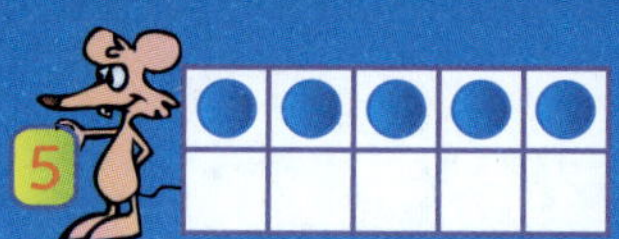

1 Colour the groups of five. Trace the numerals and the word "five".

2 Write the numbers in order.

a forwards

b backwards

3 Write 2, 0, 5 and 4 in order, smallest to largest.

4 Write 3, 5, 0 and 2 in order, smallest to largest.

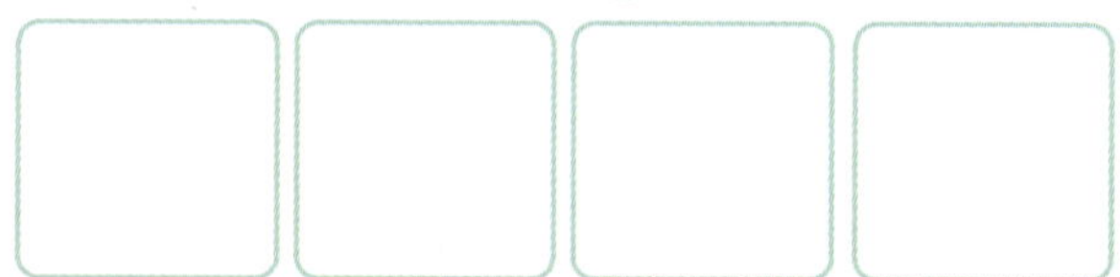

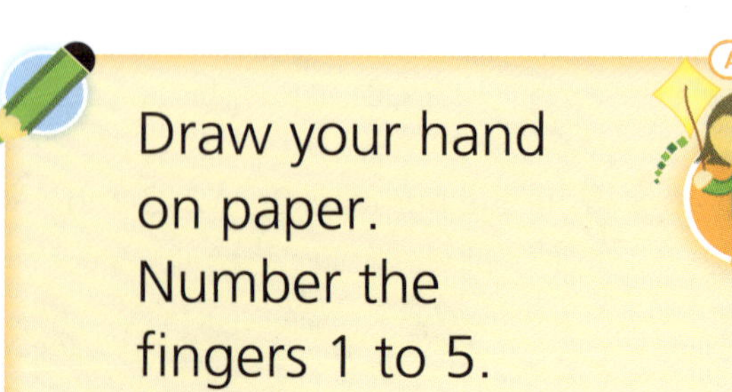

1 Colour each group differently.

How have you sorted the objects?

Which group has the most?

Sort groups of objects in your classroom. Talk about how you sorted the objects.

3A Numbers to five

1 Trace the numerals.

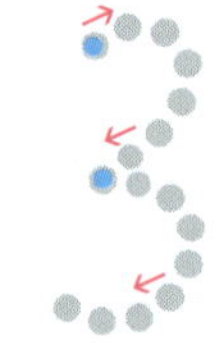
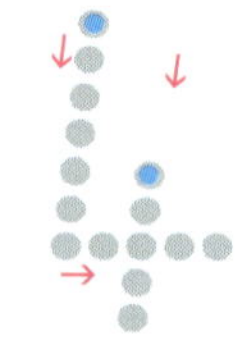
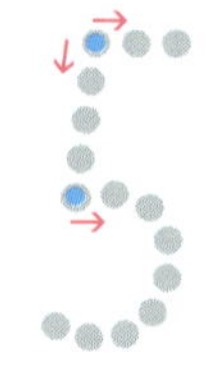

2 How many?

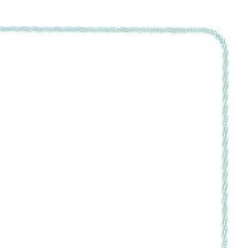

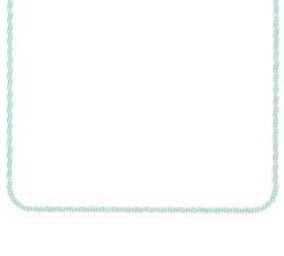

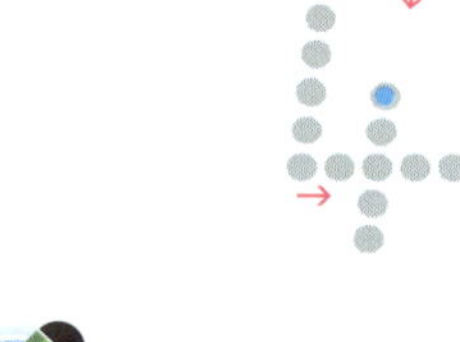

3 Join the numbers in order. Join the words in order.

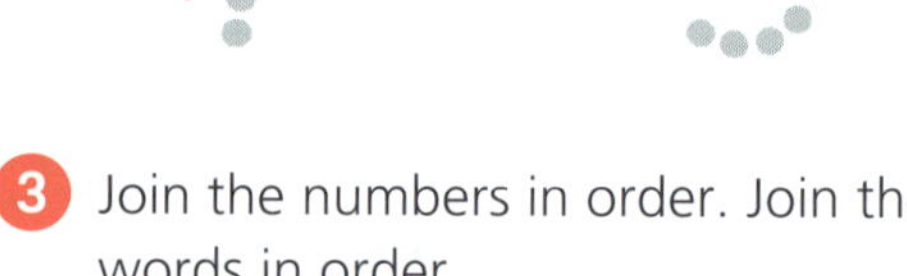
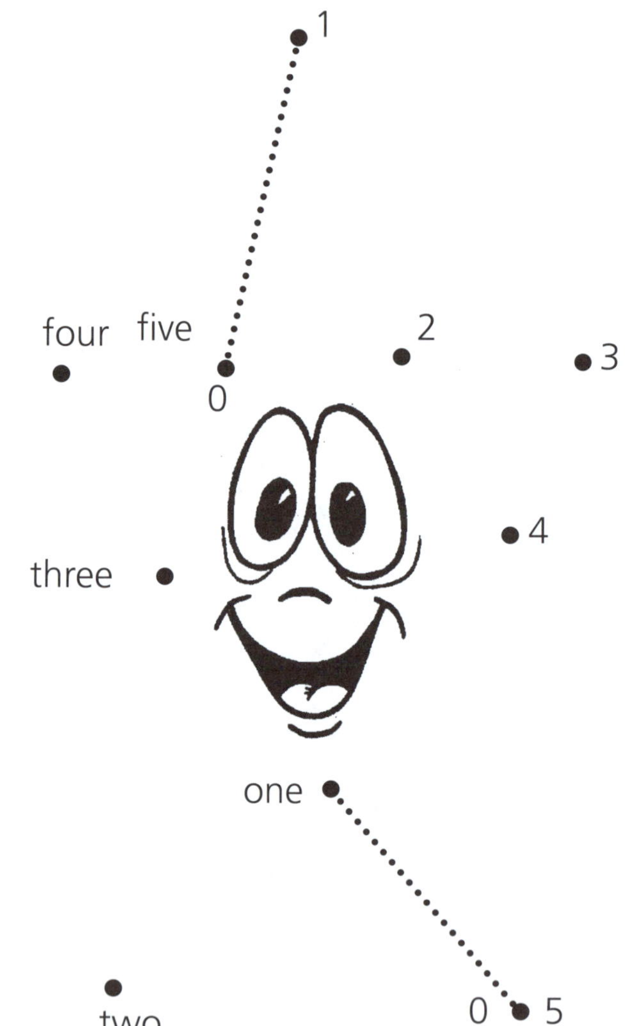

Give each person 2 legs. Give each animal 4 legs.

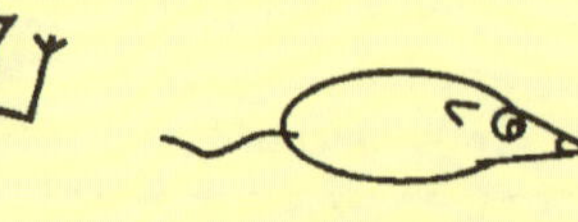

FUN SPOT

Counting to five

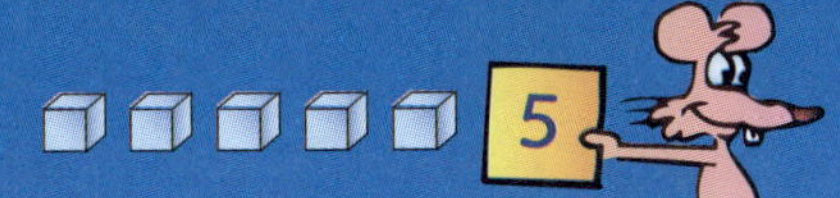

1 Trace the numerals.

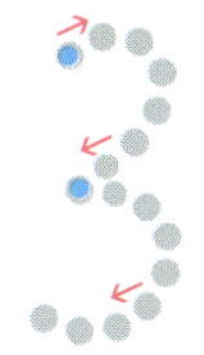
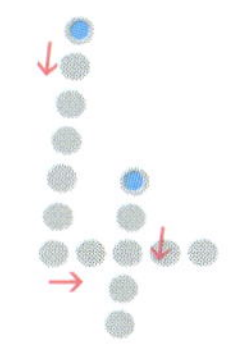
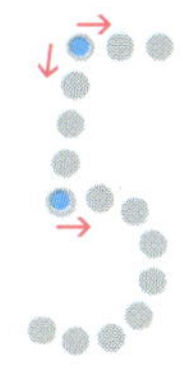

2 Talk about the picture and answer the questions.

How many ?

How many ?

How many ?

How many ?

How many ?

How many ?

How many ?

How many ?

3C The number six

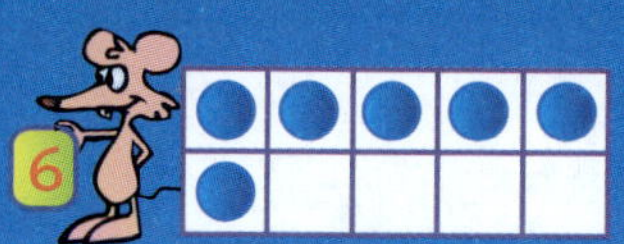

1 Colour the groups that show six. Trace the numerals and the word "six".

2 Write the number before (one less).

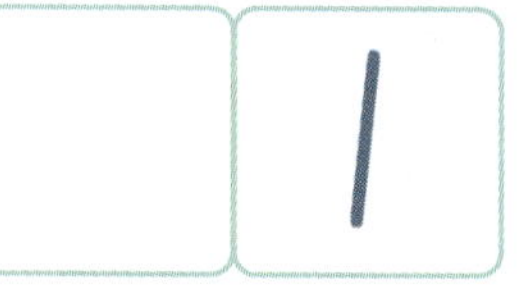

3 Write the number after (one more).

4 Draw six legs on each beetle.

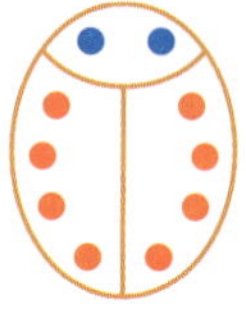

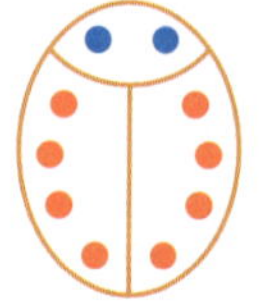

Challenge a friend

ACTIVITY

- What is one more than: 2, 4, 5, 3, 1?
- What is one less than: 3, 6, 4, 2, 5?

Egg carton game

FUN SPOT

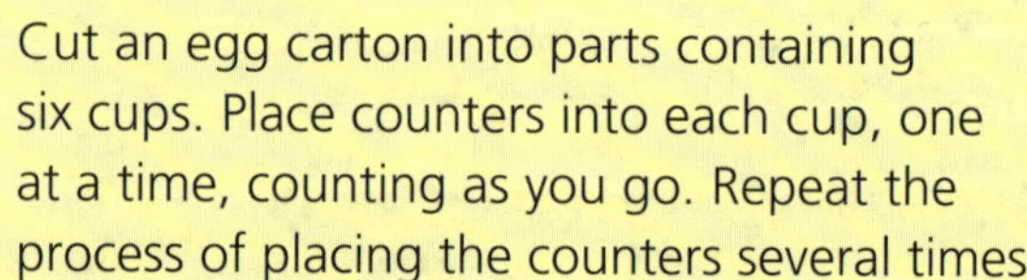

Cut an egg carton into parts containing six cups. Place counters into each cup, one at a time, counting as you go. Repeat the process of placing the counters several times.

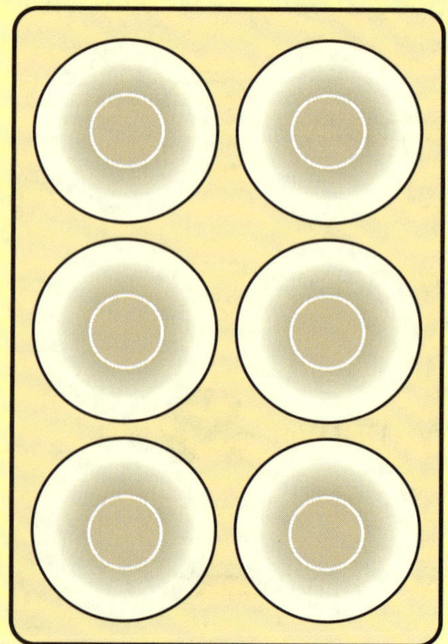

3D Curved and straight

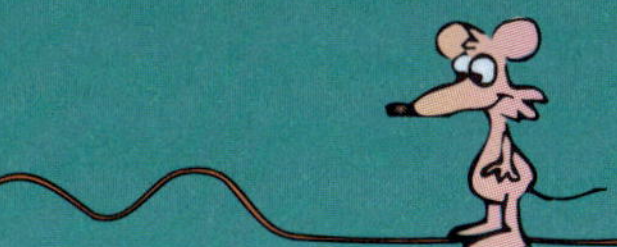

CONCEPT

- This line is curved. 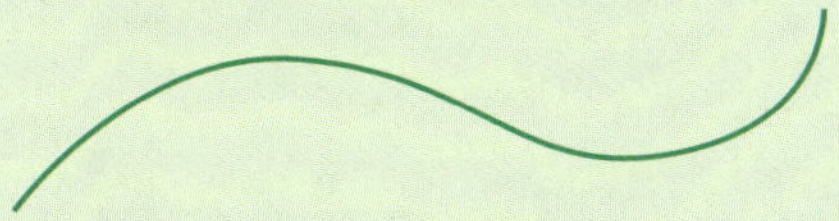
- This line is straight.

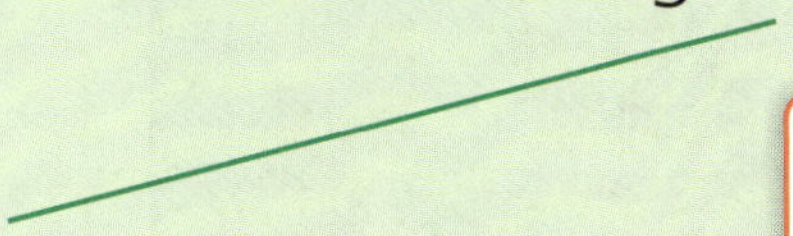

Shapes have curved or straight sides.

1 Trace the curved lines red and the straight lines blue.

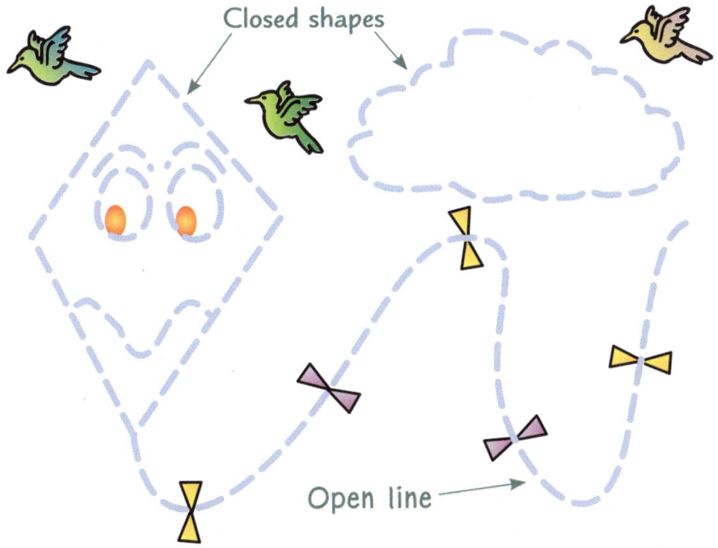

Closed shapes

Open line

2 Draw some straight and curved lines.

Straight

Curved

3 Trace these closed shapes and add some funny faces.

How many curved shapes?

How many straight shapes?

 • *AUSTRALIAN SIGNPOST MATHS NSW K* • ISBN 9780655709015

The number seven

1 Add dots to make seven. Trace the numerals and the word "seven".

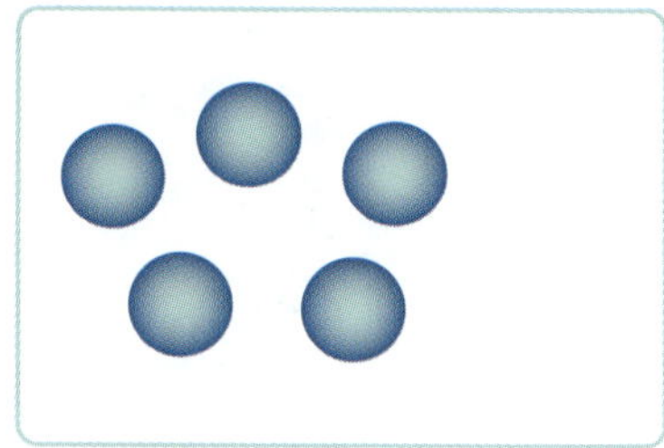

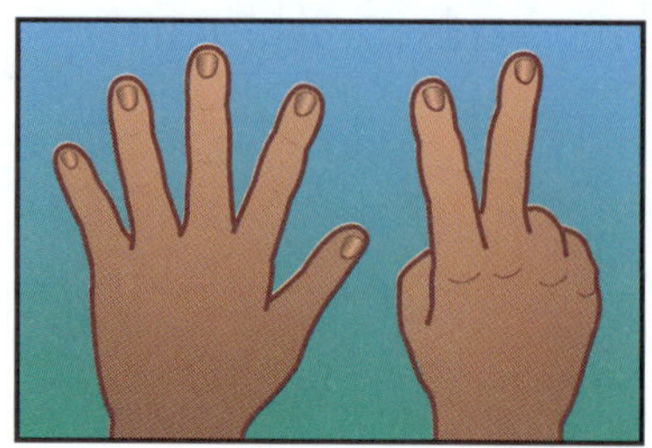

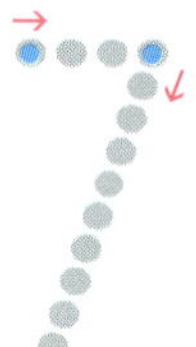

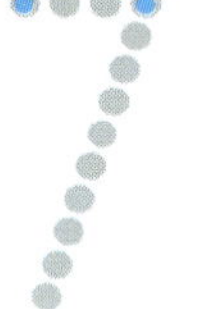

2 Write the numerals 1 to 7.

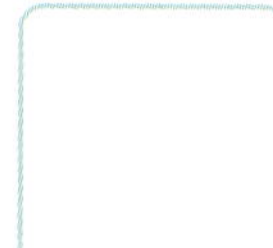

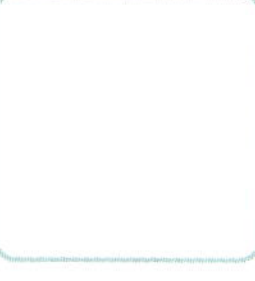

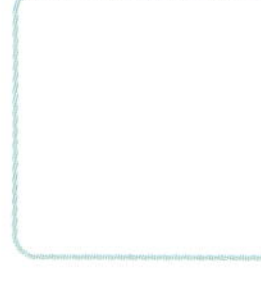

cakes

3 Draw more candles to make seven.
Draw more balloons to make seven.

Counting patterns

FUN SPOT

Lead the students in counting to seven. Use body actions, such as clapping and clicking, as each number is said.

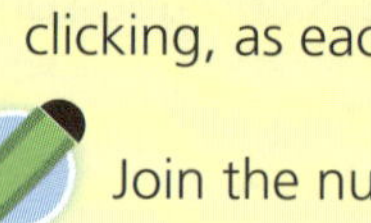

Join the numbers in order.

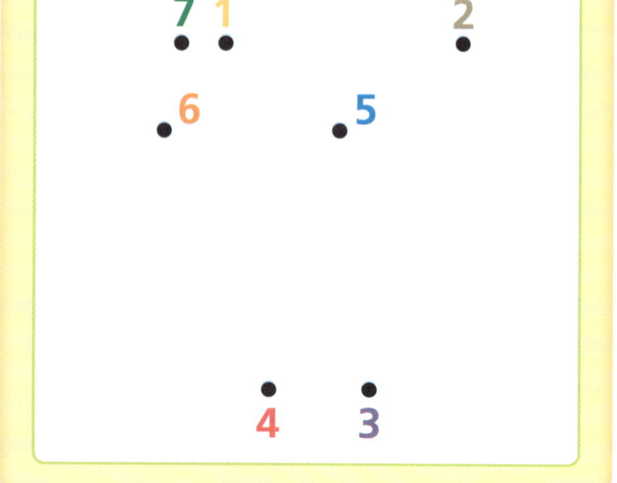

4B Dot patterns

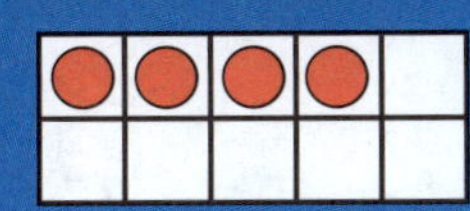
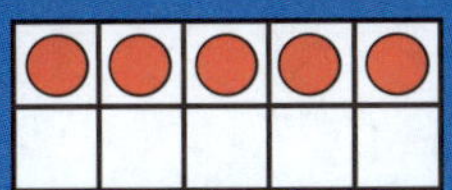
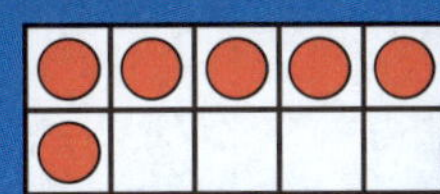

CONCEPT

Say the number of dots in each square without counting the dots.

1 In each answer square write the number of dots above it. Discuss the position of the dots in each pattern.

ACTIVITY

- Put a set of dominoes into a box. Take out one domino at a time. Say the two numbers without counting the dots.

- Keep throwing a die. ✔ Tick the boxes below when you throw:

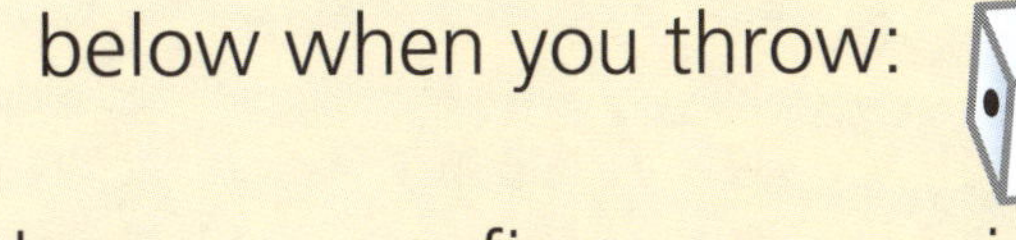
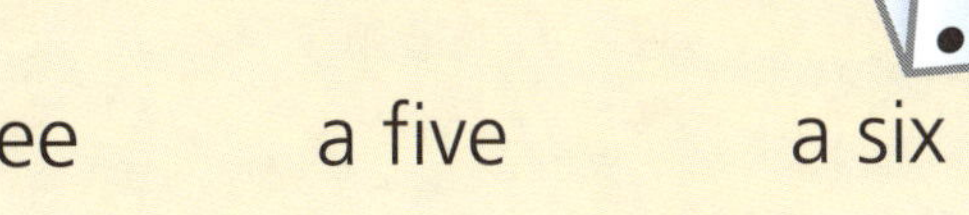

a three a five a six

4C Circles

CONCEPT

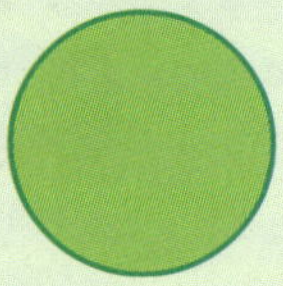

This is a circle.
It has a curved side.

1 Trace the circles in these pictures.

circle

2 Colour the circles. Discuss what makes a circle.

3 Trace the circles.

FUN SPOT

Draw a picture using circles.

This is my circle picture.

 AUSTRALIAN SIGNPOST MATHS NSW K • ISBN 9780655709015

Measurement and space

4D Heavy or light

1. Colour the light objects. **Circle** the heavy ones.

heavy

light

I am hefting.

Hefting: holding an object in each hand to feel which is heavier.

2. Compare objects by hefting. **Circle** the item that is:

heavier

lighter

5A Same and not the same

1 Draw lines to match one-to-one and then count each group.

a

the same / not the same

b

the same / not the same

c

the same / not the same

d

the same / not the same

e

the same / not the same

f

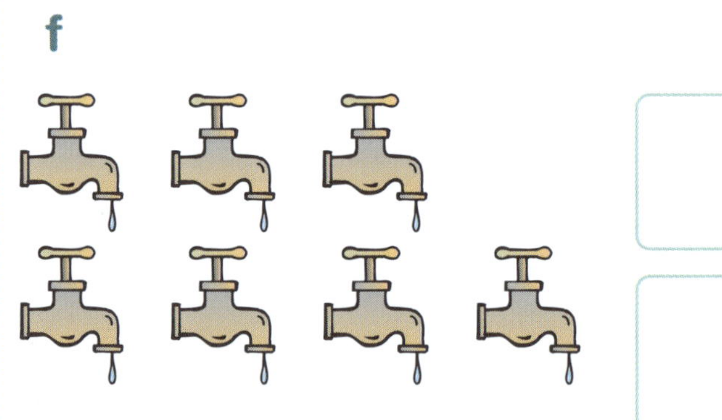

the same / not the same

2 **Circle** the parts in question 1 that show groups that are equal in size.

3 Colour the row that has more.

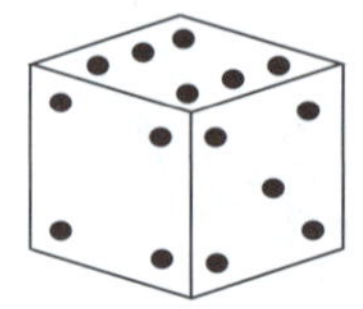
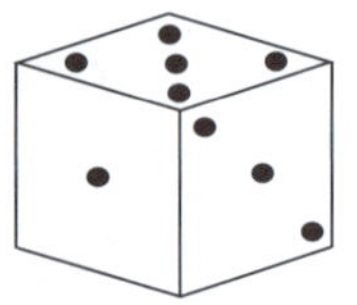
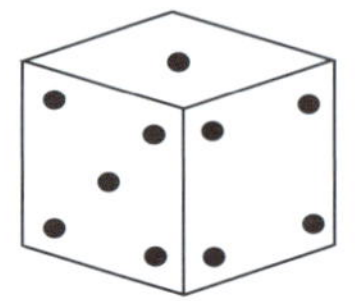

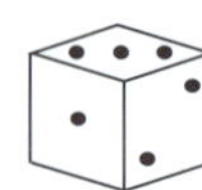

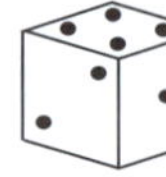
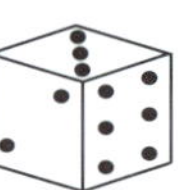
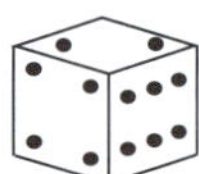

5B Same and not the same

1 Compare each row and colour the correct label.

the same as

not the same as

the same as

not the same as

2 Draw seven jellybeans in each jar so they are the **same**.

3 Draw balls in each row so that the rows are **not the same**. Write the numbers.

INVESTIGATION

Colour the groups that **are the same** size. Cross out the group that **is not the same**.

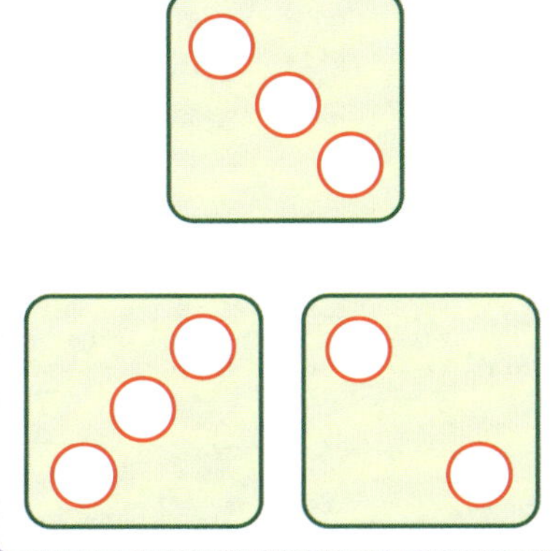

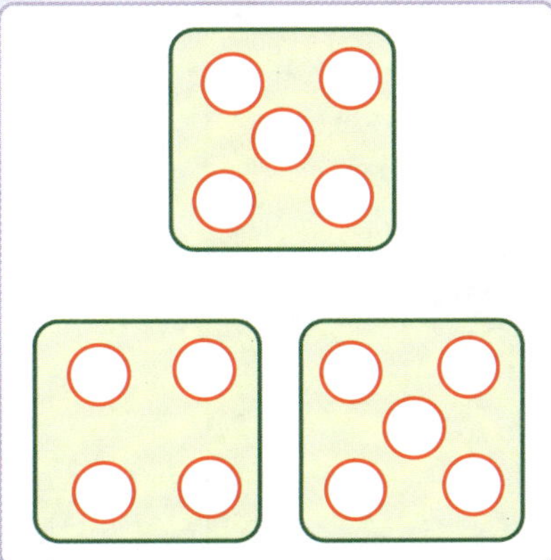

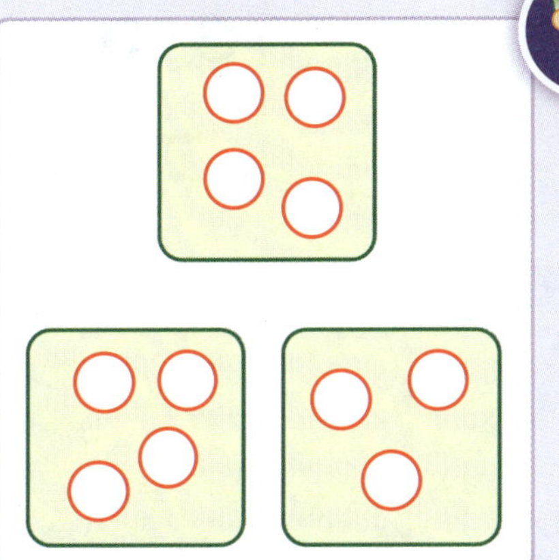

Make two groups of counters that are the same.

Make two groups of counters that are not the same.

Ask a partner if your two groups are the same or not the same.

 • *AUSTRALIAN SIGNPOST MATHS NSW K* • ISBN 9780655709015

5C Squares

CONCEPT

A square has 4 equal straight sides and 4 corners.

1 Trace the squares in these pictures.

square

2 Tick the squares. Discuss.

3 Trace the squares.

FUN SPOT

Draw a picture using squares.

This is my square picture.

 • *AUSTRALIAN SIGNPOST MATHS NSW K* • ISBN 9780655709015

Full, empty and half full

1 Circle the things that are full. Tick the things that are empty.

How many glasses are about half full? ☐ Talk about the picture.

2 Draw lollies to make the jars:

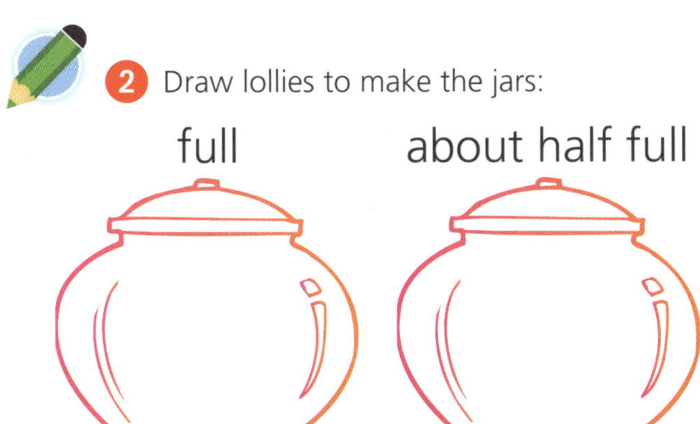

3 Colour two empty boxes.

 • *AUSTRALIAN SIGNPOST MATHS NSW K* • ISBN 9780655709015

6A The number eight

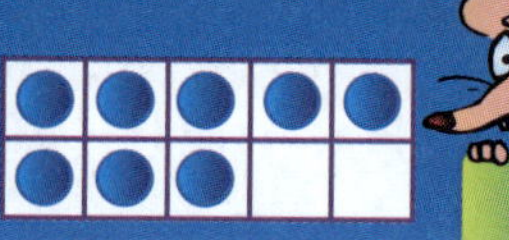

1 Trace the numerals and the word "eight".

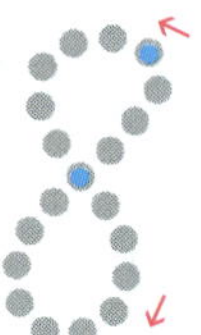
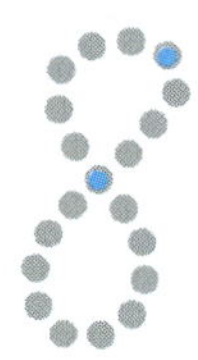
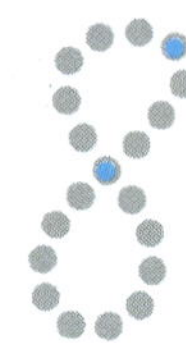
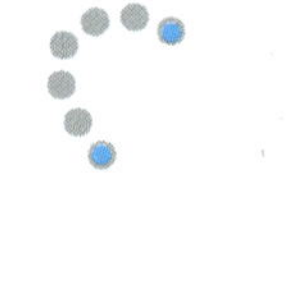

2 **Circle** boxes that show eight.

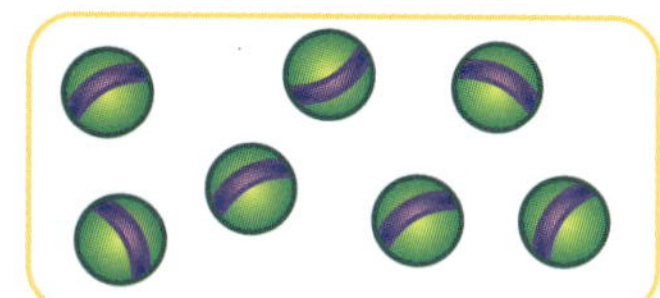

eight

3 Count on.

6		
3		

4 Count back.

8		
4		

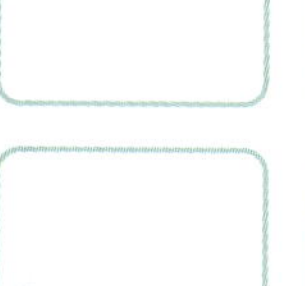

5 Draw a face in each window.

☐ faces

INVESTIGATION

Craft stick patterns

Ask students to count out eight craft sticks and use them to make two squares. Count the craft sticks used for each shape.

Draw the shapes you have made.

 ISBN 9780655709015

Comparing groups

1 a Count the objects.

b Circle to show the largest group above.

c Circle to show the smallest group above.

d Colour: 3  5 6 7

INVESTIGATION

Balloons at the party

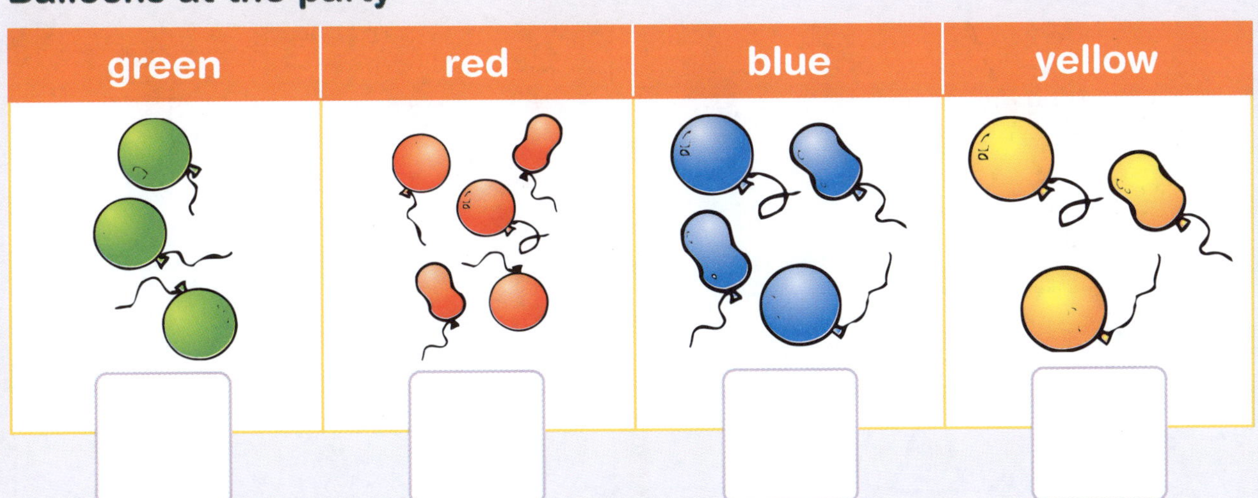

- How many balloons of each colour?
- There are more ______ balloons than blue balloons.

6C Triangles

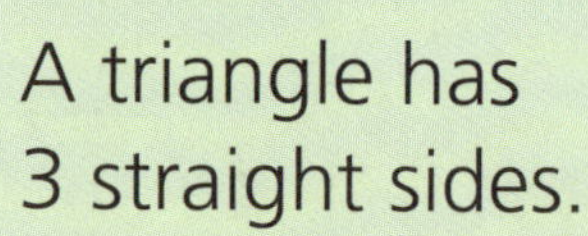

CONCEPT

A triangle has 3 straight sides.

1 Trace the triangles in these pictures.

3 Trace the triangles.

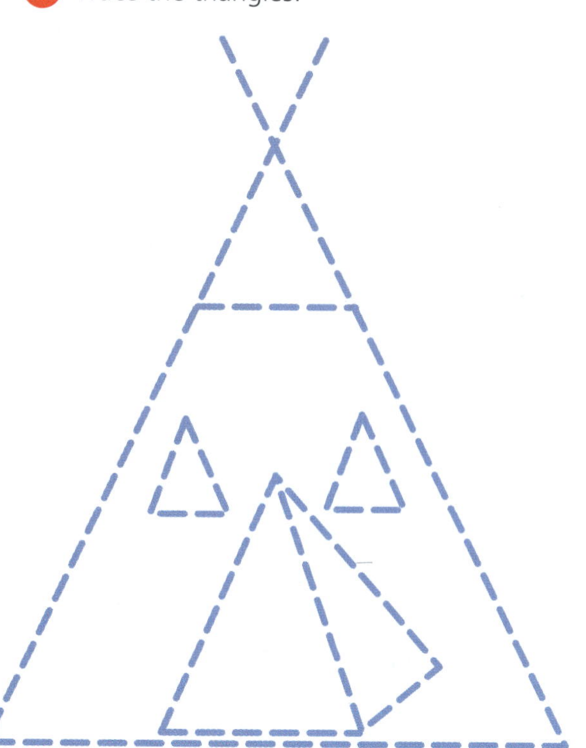

2 Colour the triangles. Discuss.

FUN SPOT

Draw a picture using triangles.

This is my triangle picture.

6D Comparison of mass

CONCEPT

Holding an object in each hand helps you decide which one is heavier. This is called hefting.

1 Colour the light objects.

ACTIVITY

Predict which object is heavier or lighter. Compare objects by hefting.
Circle the lightest object. ✘ Cross the heaviest object.

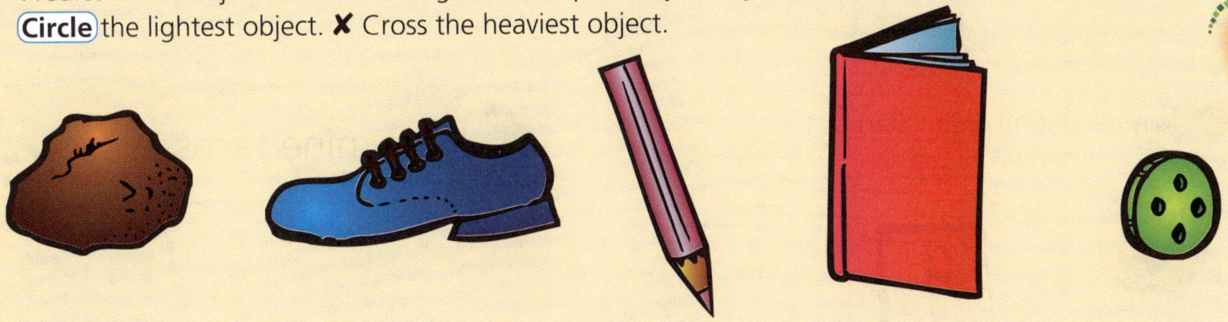

rock	shoe	pencil	book	button

________ is heavier than ________.

________ is lighter than ________.

________ is heavier than ________.

 • *AUSTRALIAN SIGNPOST MATHS NSW K* • ISBN 9780655709015

7A The number nine

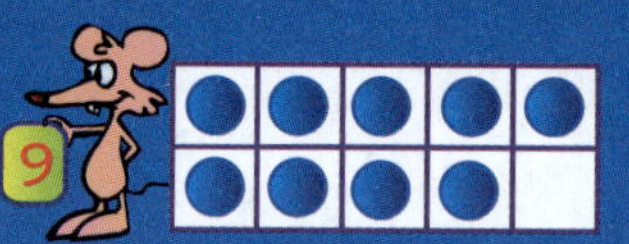

1 Draw a dot pattern for nine. Trace the numerals and the word "nine".

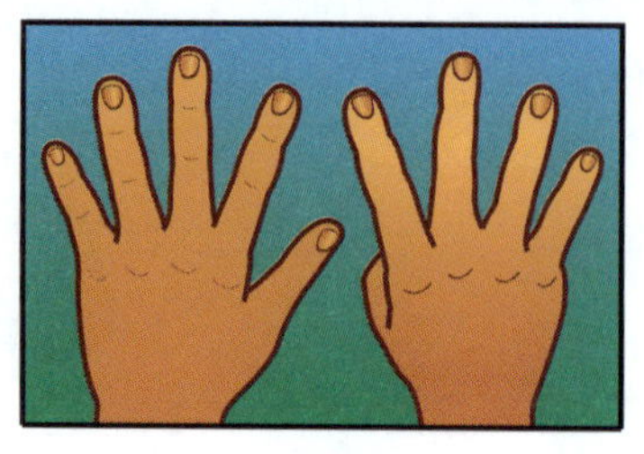

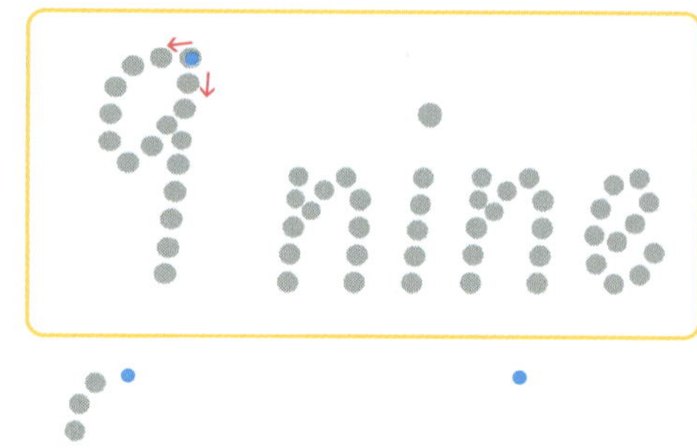

2 Write how many dots you can see in each domino altogether.

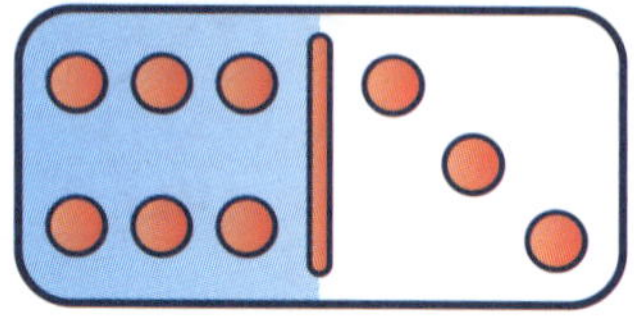

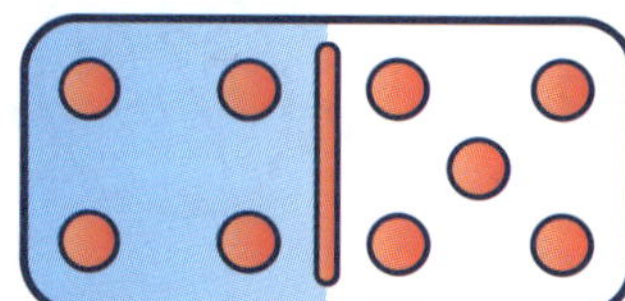

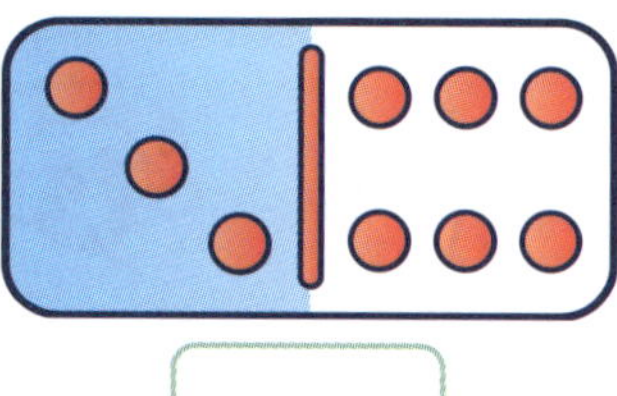

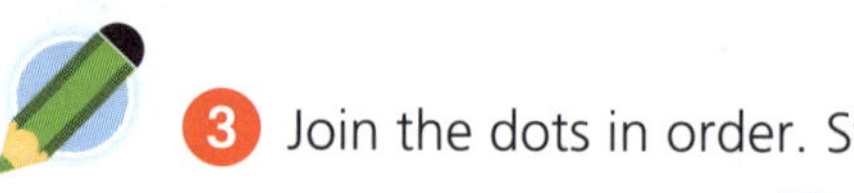

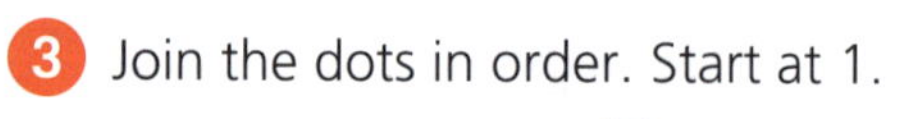

3 Join the dots in order. Start at 1.

How many ants?

Draw nine faces.

Go outside

Ask students to skip, bounce a ball or hop while the rest of the class counts.

 • *AUSTRALIAN SIGNPOST MATHS NSW K* • ISBN 9780655709015

The number ten

1 Draw a dot pattern for 10. Trace the numerals and the word "ten".

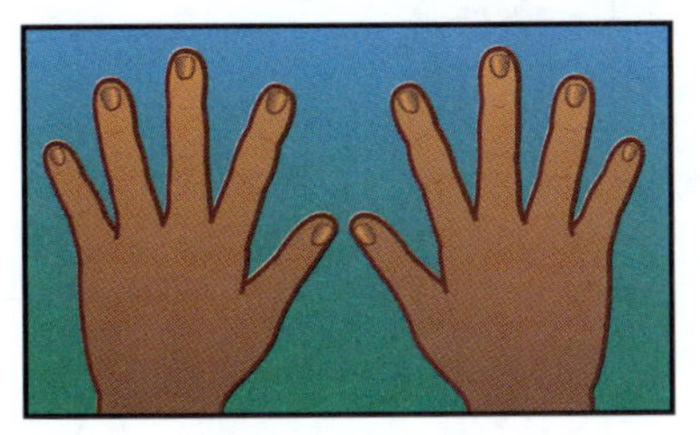
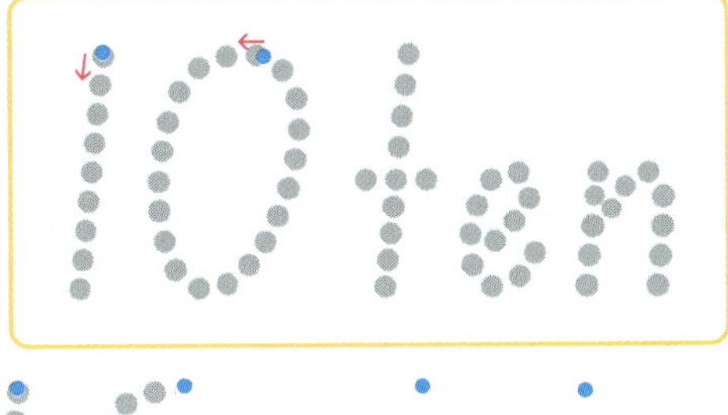

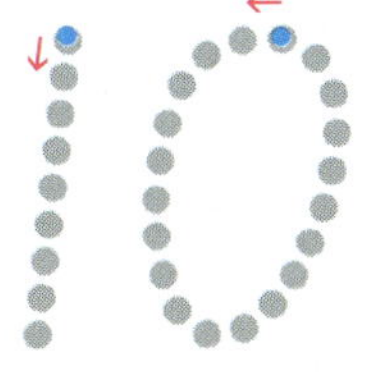
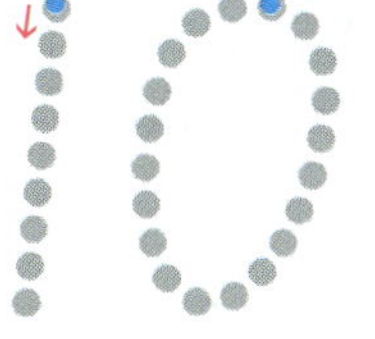

2 **Circle** the cards that have 10 pictures.

3 Order the cards from smallest to largest.

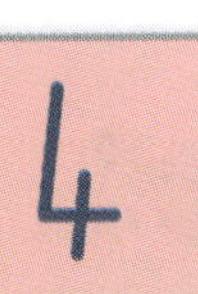

5

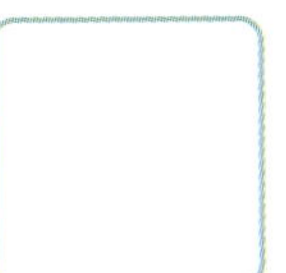
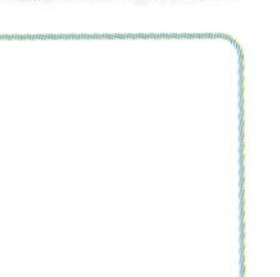

4 How many fingers on two hands?

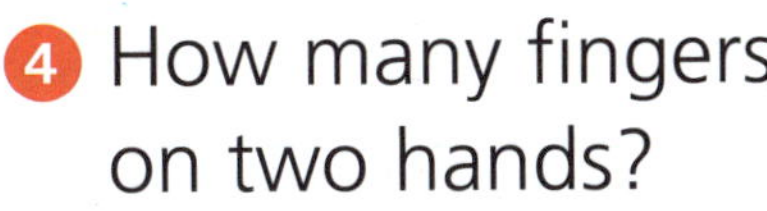
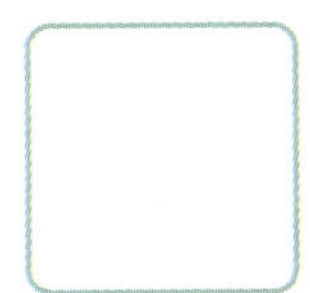

FUN SPOT

Concentration

Make two sets of 6–10 number cards, as shown. Place both sets of cards face down on a table. Students take turns to turn over two cards. If the cards match, the student keeps them.

Match these cards.

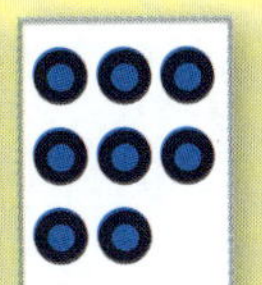

7C Rectangles

CONCEPT

A rectangle has two equal long sides and two equal short sides, like a stretched square.

1 Trace the rectangles in these pictures.

2 Put patterns on the rectangles. Discuss.

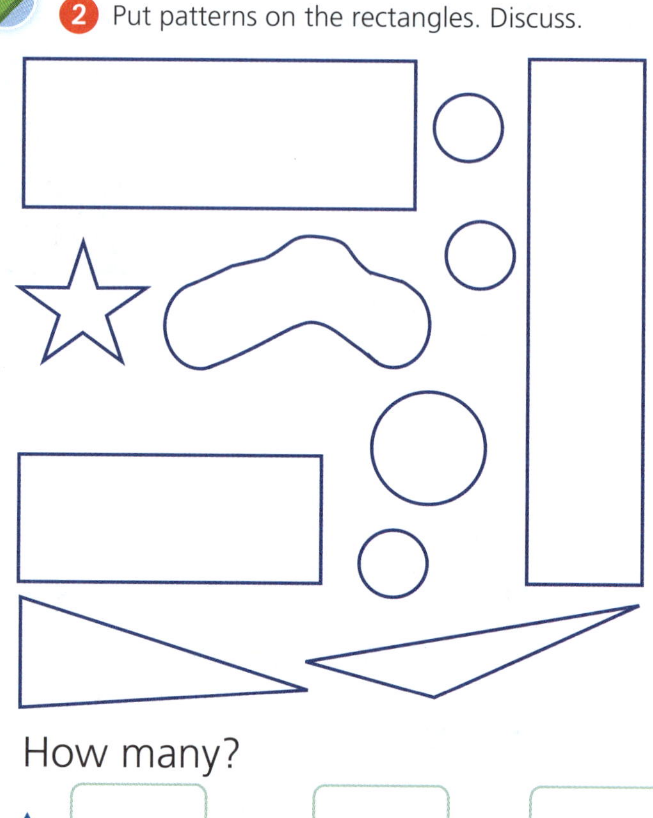

How many?

3 Trace the rectangles.

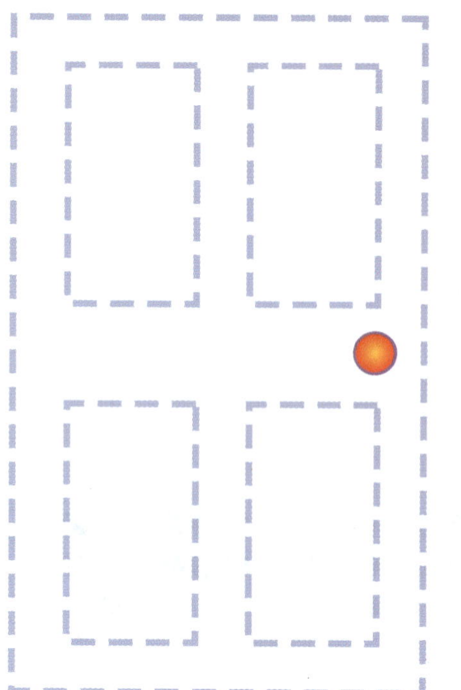

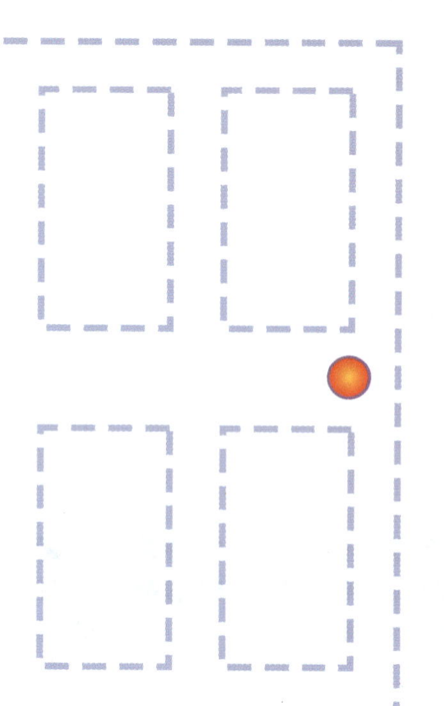

FUN SPOT

Draw a picture using rectangles.

This is my rectangle picture.

7D Cutting shapes

ACTIVITY

- Cut and paste a circle.

circle

- Cut and paste a square.

square

- Cut and paste a triangle.

triangle

- Cut and paste a rectangle.

rectangle

© PEARSON AUSTRALIA 2023 • *AUSTRALIAN SIGNPOST MATHS NSW K* • ISBN 9780655709015

8A Numbers to ten

10 is 1 ten
and 0 ones.

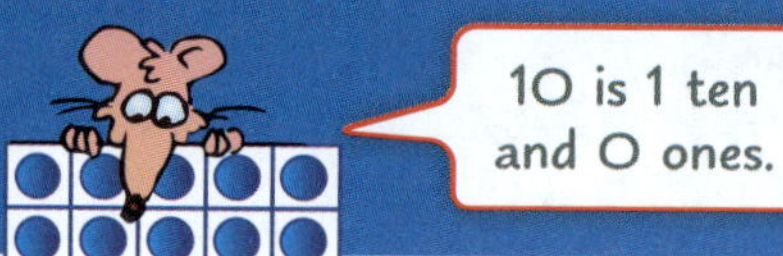

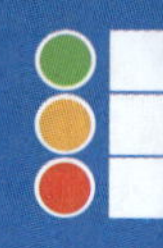

1 Write the numerals on the rocket. Match the words and numerals. Trace the numerals and words.

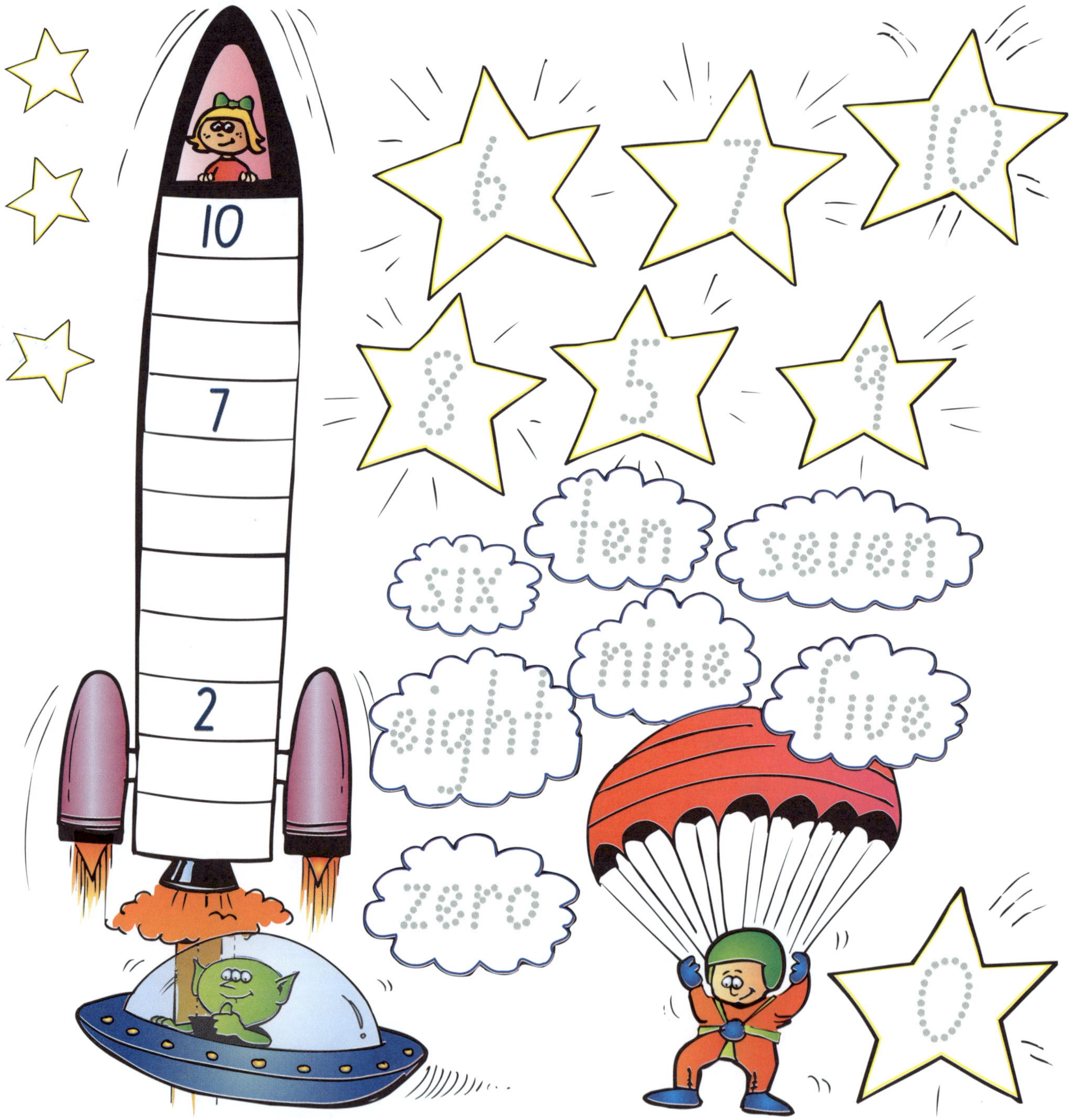
10
7
2
6
7
10
8
5
9
ten
seven
six
nine
eight
five
zero
0

2 Talk about the picture above and answer the questions.

How many?

 • *AUSTRALIAN SIGNPOST MATHS NSW K* • ISBN 9780655709015

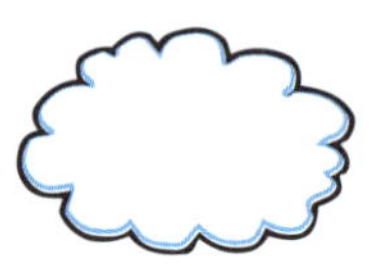

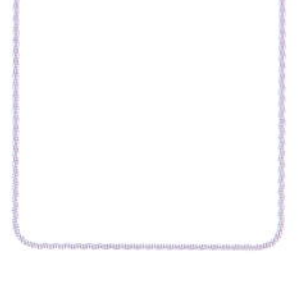

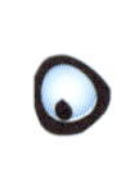

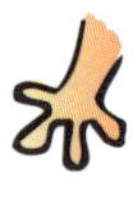

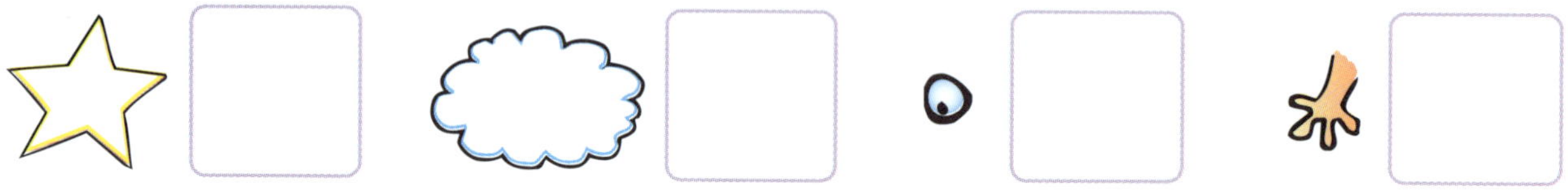

8B Numbers to ten

1 Talk about the picture and answer the questions.

How many ?

How many ?

How many ?

How many ?

How many ?

How many ?

2 Count forwards.

2

6

3 Count backwards.

8

10

Draw dots to make 10 on each domino.

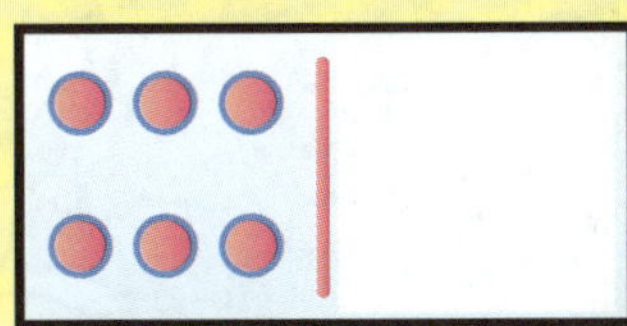

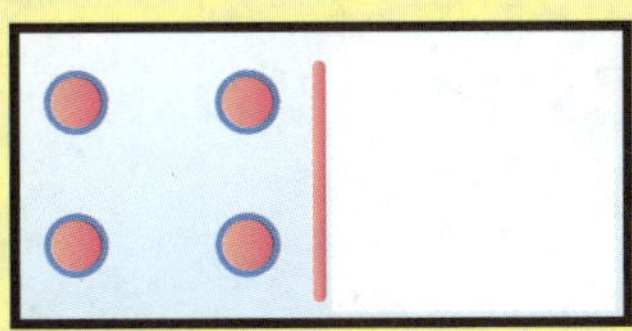

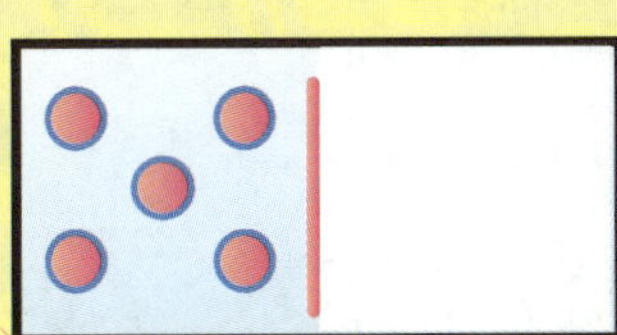

- Count all of the dots already drawn in red.
- See if you can count to 20.

8C Position

FUN SPOT

Colour what is:

- on the log black
- in front of the tree pink
- inside the tent purple
- near the fire orange
- under the log red
- close to the log brown
- behind the tent green
- up on a tree branch orange
- beside the mouse yellow.

Talk about the positions of the other objects.

Language of location

1 Draw:

 on top of

 inside

 outside

 behind

 between

 under

 in

 in front of

next to

2 Draw a circle behind the boy, a square next to him, and a triangle in front of him.

3 Write up or down.

 AUSTRALIAN SIGNPOST MATHS NSW K • ISBN 9780655709015

9A Numbers to 10

1 Write the numbers up to 10 on the clothes. Draw ten flowers under the washing line.

2 How many dots?

Draw other patterns using 10 dots.

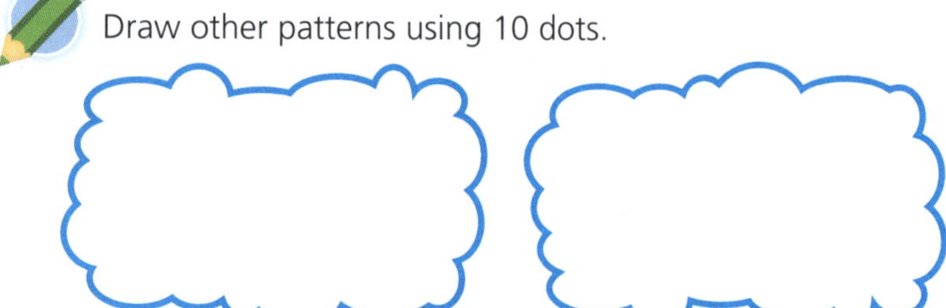

3 Write the number that comes:

before		after
	9	
	4	
	6	
	7	
	5	

4

How many legs?

5

How many grey circles?

How many black circles?

6 See if you can count to 30.

 • *AUSTRALIAN SIGNPOST MATHS NSW K* • ISBN 9780655709015

9B Numbers 11 and 12

1 Trace the numerals and words below.

2 Count and write the number of objects.

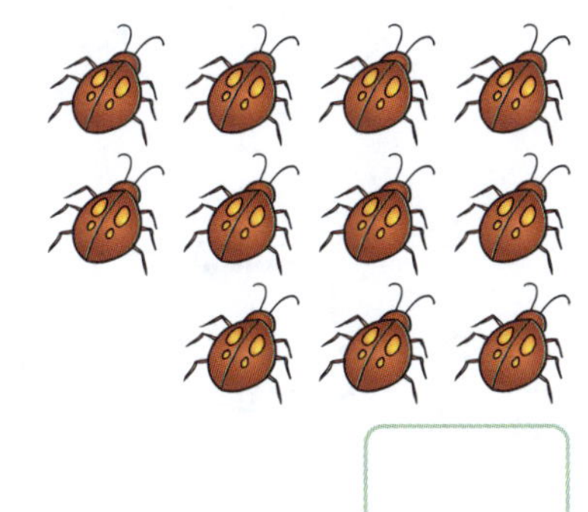

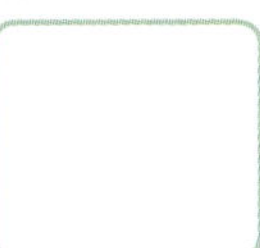

3 Draw 12 circles. (○)

Draw 11 stars. (∗)

4 Count forwards and backwards. Discuss numbers "before" and "after" a given number.

1	2	3	4	5	6	7	8	9	10
11	12	13	14	15	16	17	18	19	20

 • *AUSTRALIAN SIGNPOST MATHS NSW K* • ISBN 9780655709015

Longer and shorter

1 Length is the measure of an object from end to end.
Colour the longer things. **Circle** the shorter things.

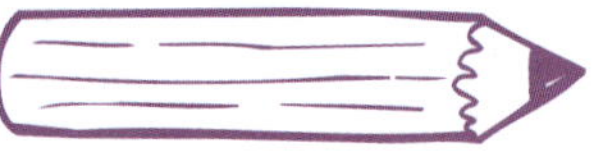

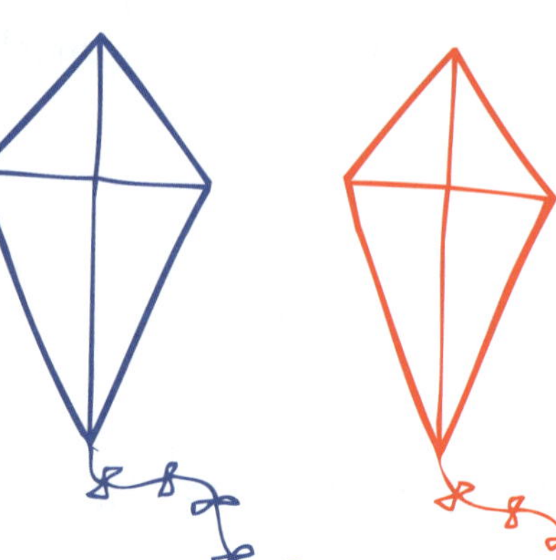

longer

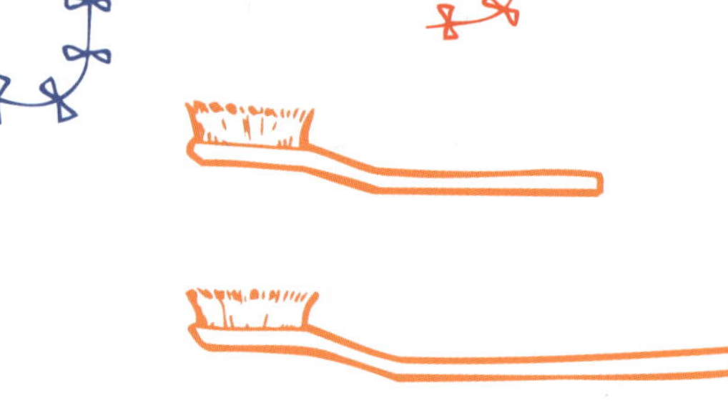

shorter

→ Tell a friend how you know which object is longer.

Use three coloured pencils of different length.
Line them up end-to-end at the dots on the page.
Put a dot at the end of each pencil. Draw a line
for each pencil. Write what you found.

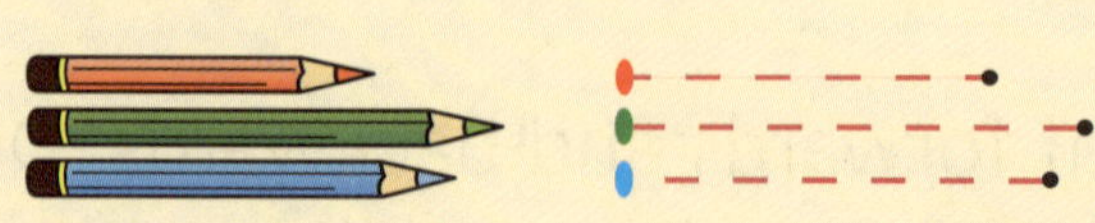

The ______ pencil is longer than the ______ pencil.

Daytime and night-time

1 Draw something you do when it's daytime.

2 Draw something you do when it's night-time.

8 o'clock.

3 Colour things that you do in the morning. **Circle** the event that takes the longest time. Discuss the order of these events.

10A Adding two groups

1 Draw a picture to show how many altogether.

a  and  makes 

b  and  makes 

c 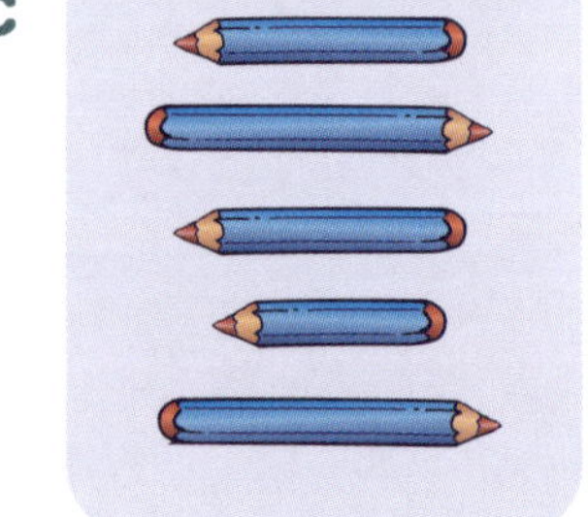 and 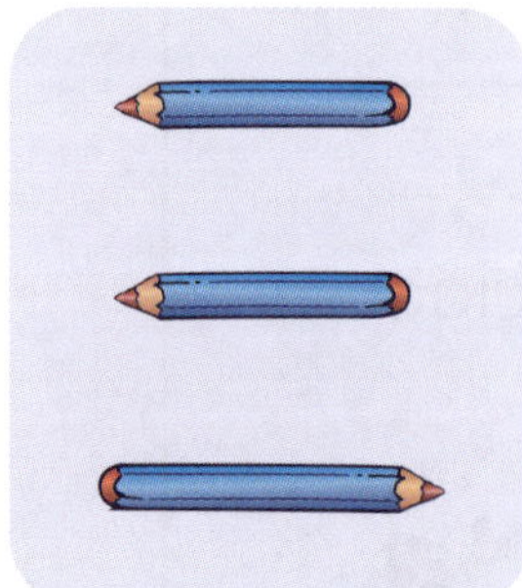 makes 

2

☐ fish and  fish makes  fish altogether.

 • *AUSTRALIAN SIGNPOST MATHS NSW K* • ISBN 9780655709015

Adding two groups

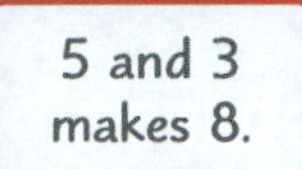

1 Draw a picture to show how many altogether.

 and  makes

 and  makes

 and  makes

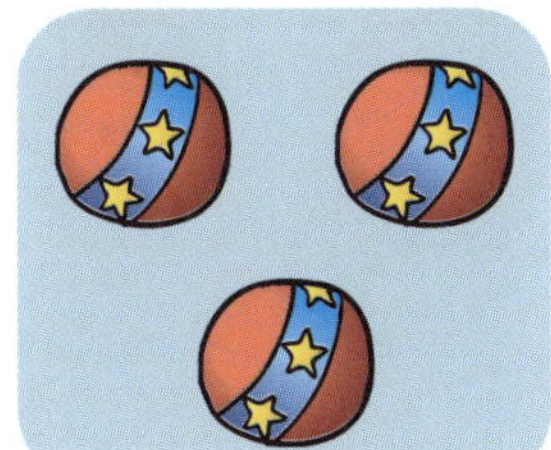 and 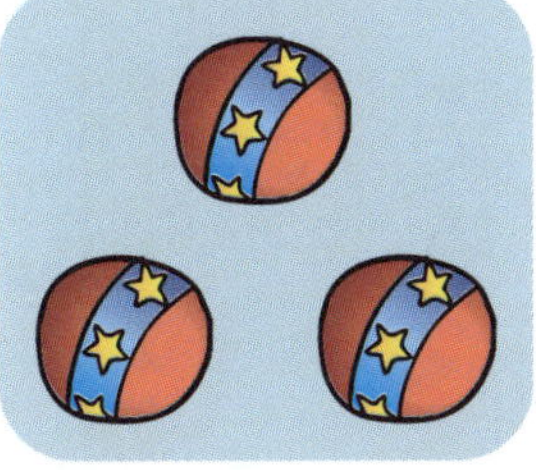 makes

2 Draw:

3 apples

4 apples

Explain what you have done.

a How many altogether?

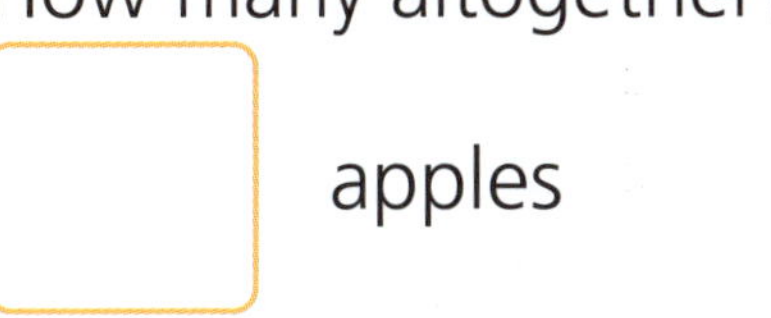

apples

b 3 apples and 4 apples

makes ☐ apples.

10C Shape pictures

1 Discuss the picture. Colour the shapes at the side with the same colours that are used in the picture.

2 Use circles, triangles and rectangles to make a flower.

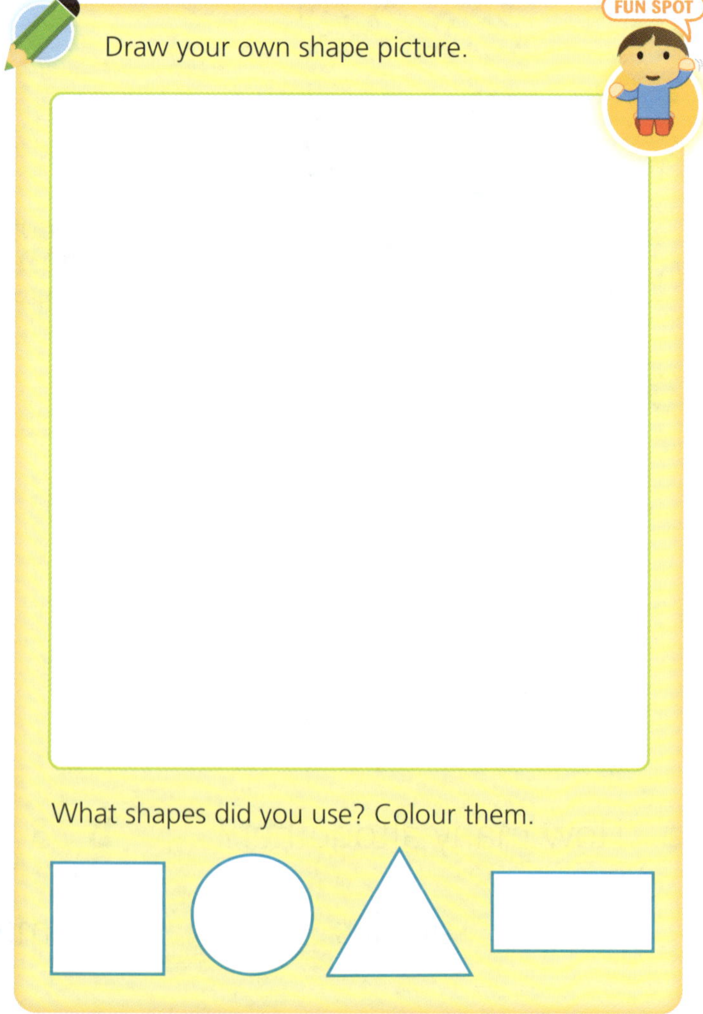

10D Morning and afternoon

1 Colour the correct label and trace the words. Discuss daily events.

morning	morning	morning
afternoon	afternoon	afternoon

2 Draw something that you do in the morning.

morning

3 Draw something that you do in the afternoon.

afternoon

4 Put these events in order, writing 1 to 4 beside them.

dinner ☐ get out of bed ☐ go to bed ☐ breakfast ☐

 • *AUSTRALIAN SIGNPOST MATHS NSW K* • ISBN 9780655709015

11A Numbers 13 to 20

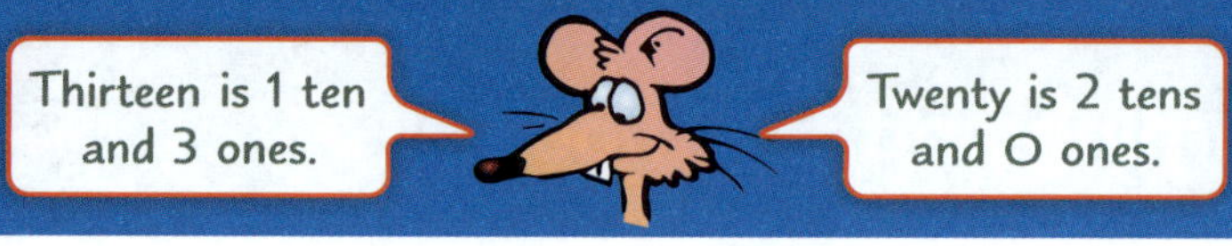

Thirteen is 1 ten and 3 ones.

Twenty is 2 tens and 0 ones.

1 Trace over the numerals and words.

13 thirteen

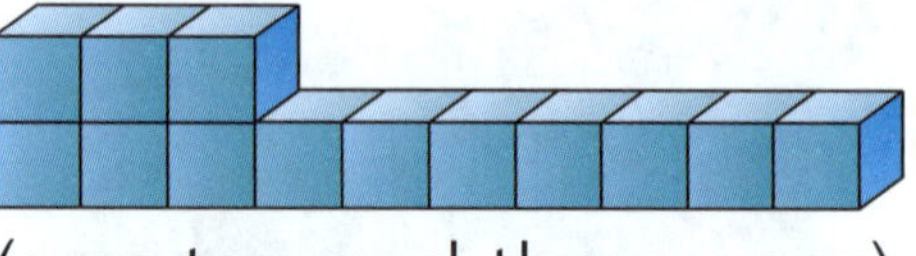

(one ten and three ones)

14 fourteen

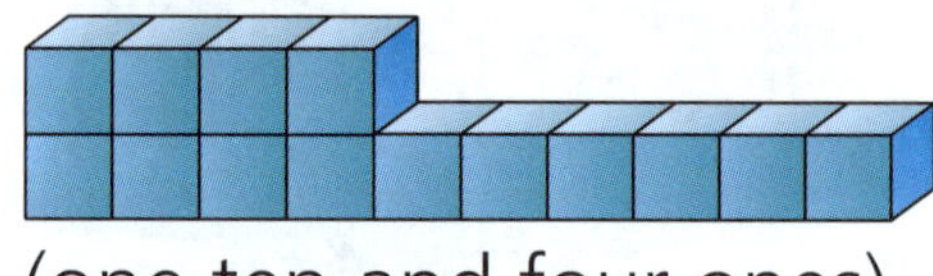

(one ten and four ones)

15 fifteen

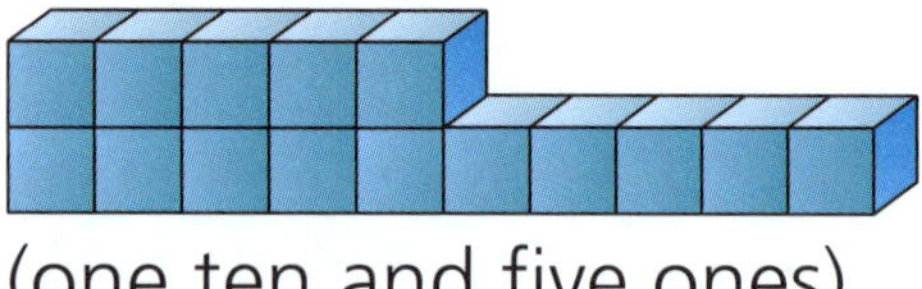

(one ten and five ones)

16 sixteen

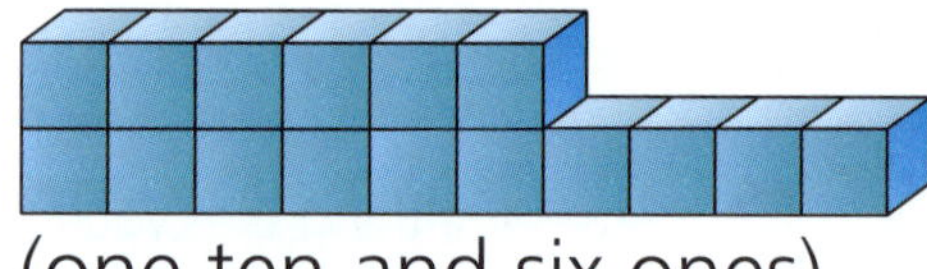

(one ten and six ones)

17 seventeen

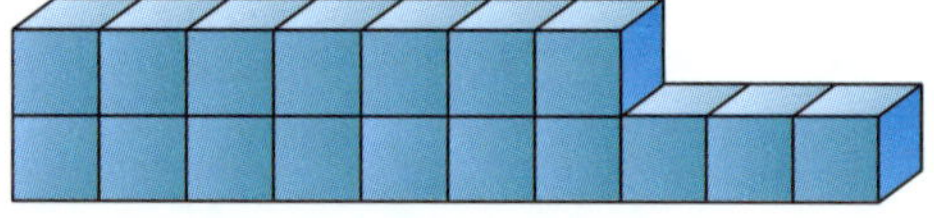

(one ten and seven ones)

18 eighteen

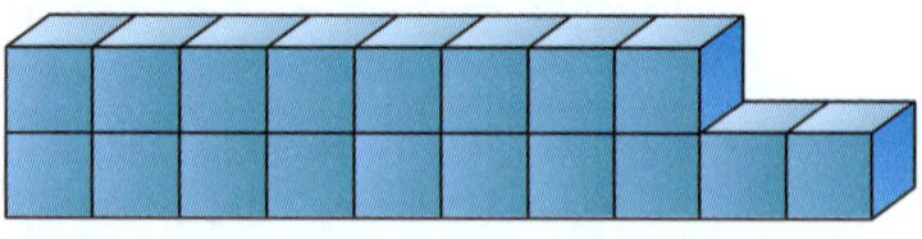

(one ten and eight ones)

19 nineteen

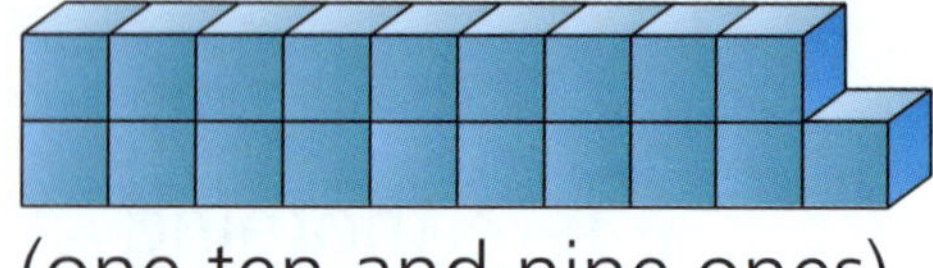

(one ten and nine ones)

20 twenty

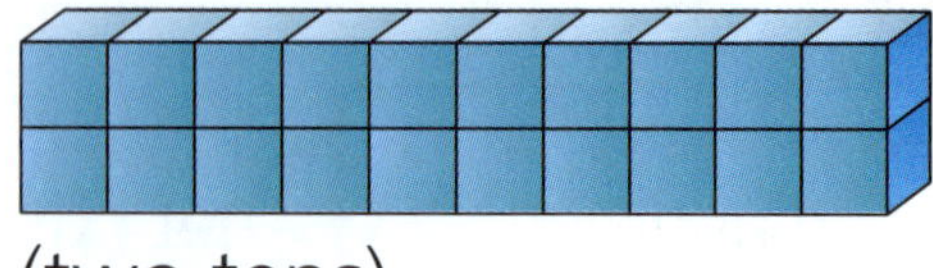

(two tens)

11B Numbers 11 to 20

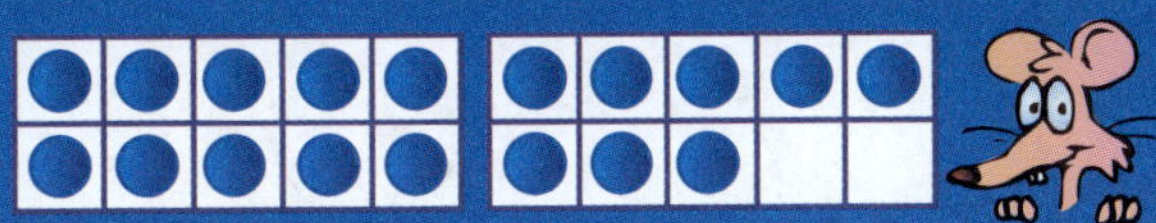

1 Complete the number pattern and say each number as tens and ones. 14 is 1 ten and 4 ones.

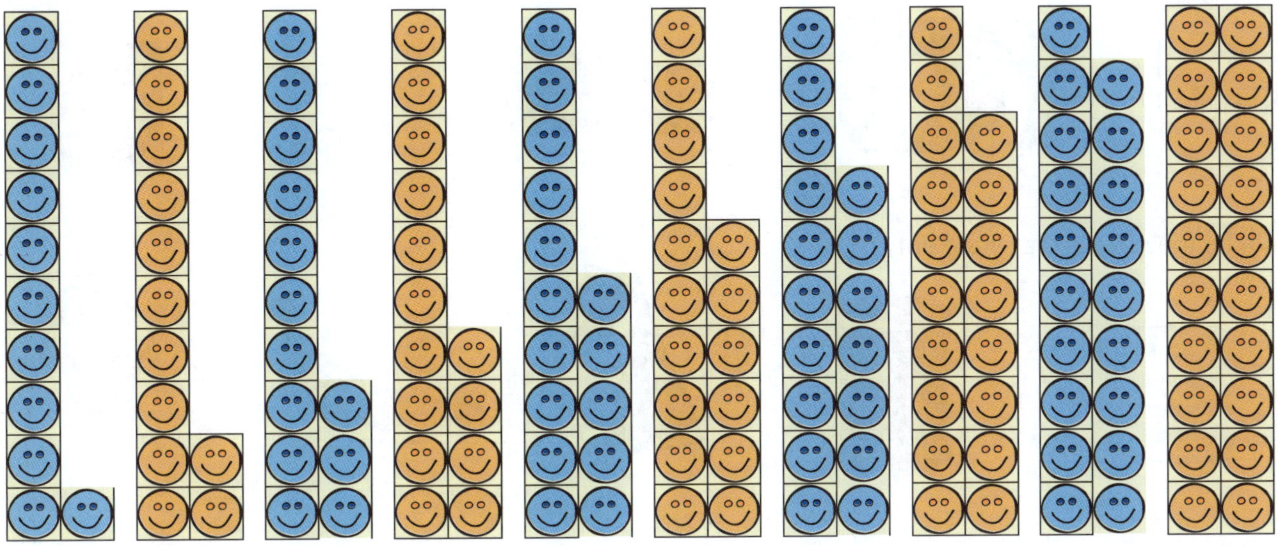

11	12								20

The pattern rule is add ______.

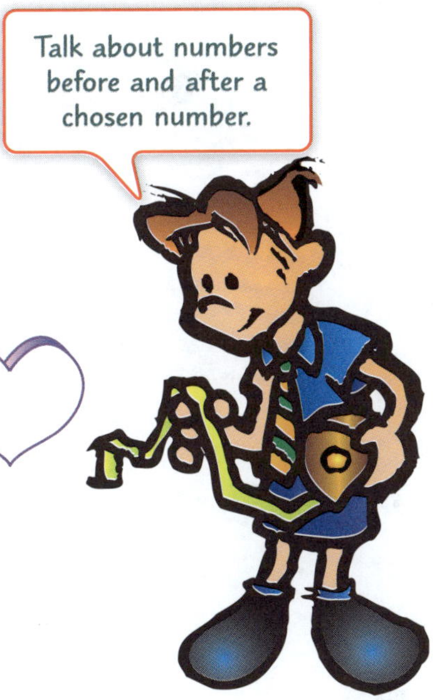

2 Complete the number patterns and talk about them.

13, 14, 15, ___, ___, ___, ___

20, 19, 18, ___, ___, ___

3 Make a list of the 'teen' numbers.

___, ___, ___, ___, ___, ___, ___

4 Put these numbers in order: 19, 16, 18, 15, 13, 17, 14.

___, ___, ___, ___, ___, ___, ___

11C Sorting objects

1 **Circle** the objects that are curved.

2 **Circle** the objects that are pointy.

3 **Circle** the objects that have flat surfaces.

4 Colour the boxes that describe each object.

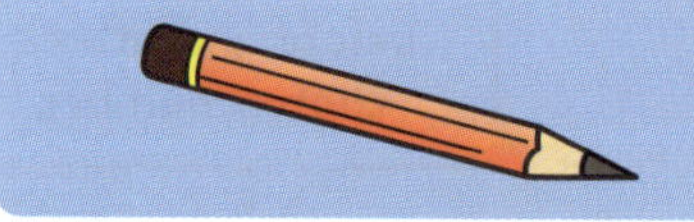		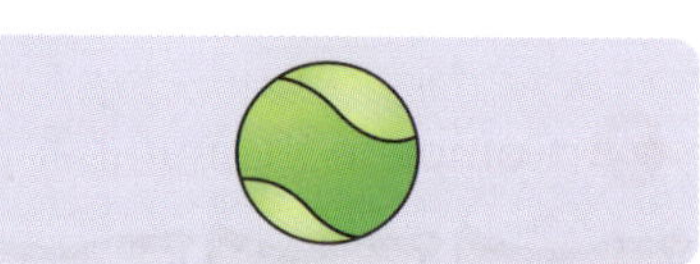
flat	flat	flat
round	round	round
pointy	pointy	pointy
curved	curved	curved
straight	straight	straight

5 Draw objects that will stack. In pairs, discuss reasons for your choice.

 • *AUSTRALIAN SIGNPOST MATHS NSW K* • ISBN 9780655709015

3D objects

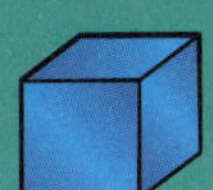

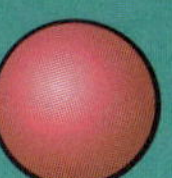

1 Pam found objects the same shape as the yellow blocks. Describe and name each object, then match each block to the correct object. Tick the objects that can stack.

2 Trace each block. Draw lines from each word to the matching blocks.

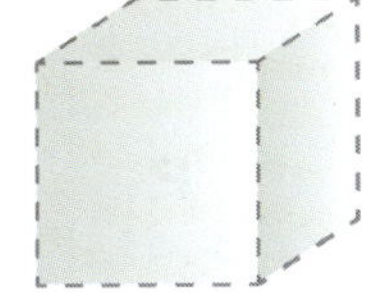

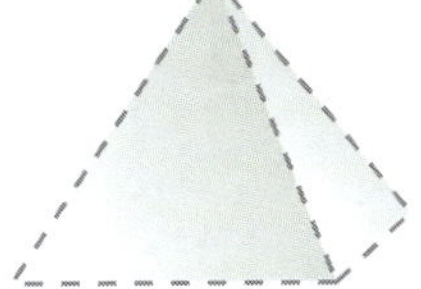

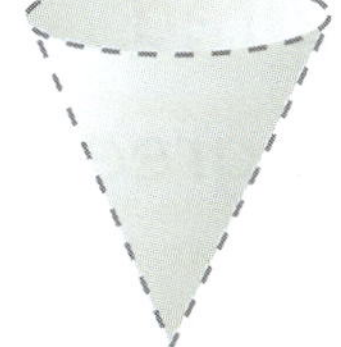
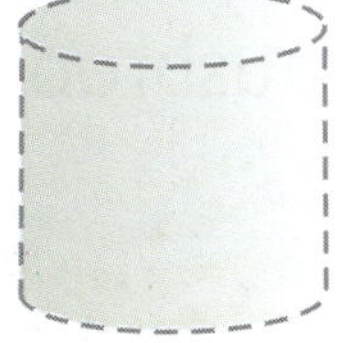

curved

straight

pointy

round

flat

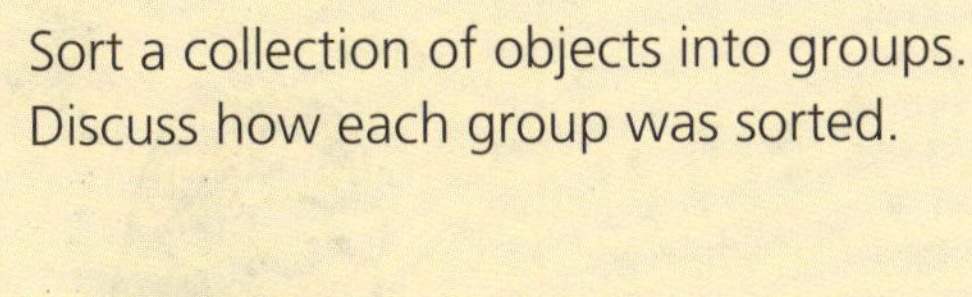

ACTIVITY

Sort a collection of objects into groups.
Discuss how each group was sorted.

12A Addition

CONCEPT

We can combine 2 groups to make a larger one.

2 and 3 makes 5.

1 Count how many altogether.

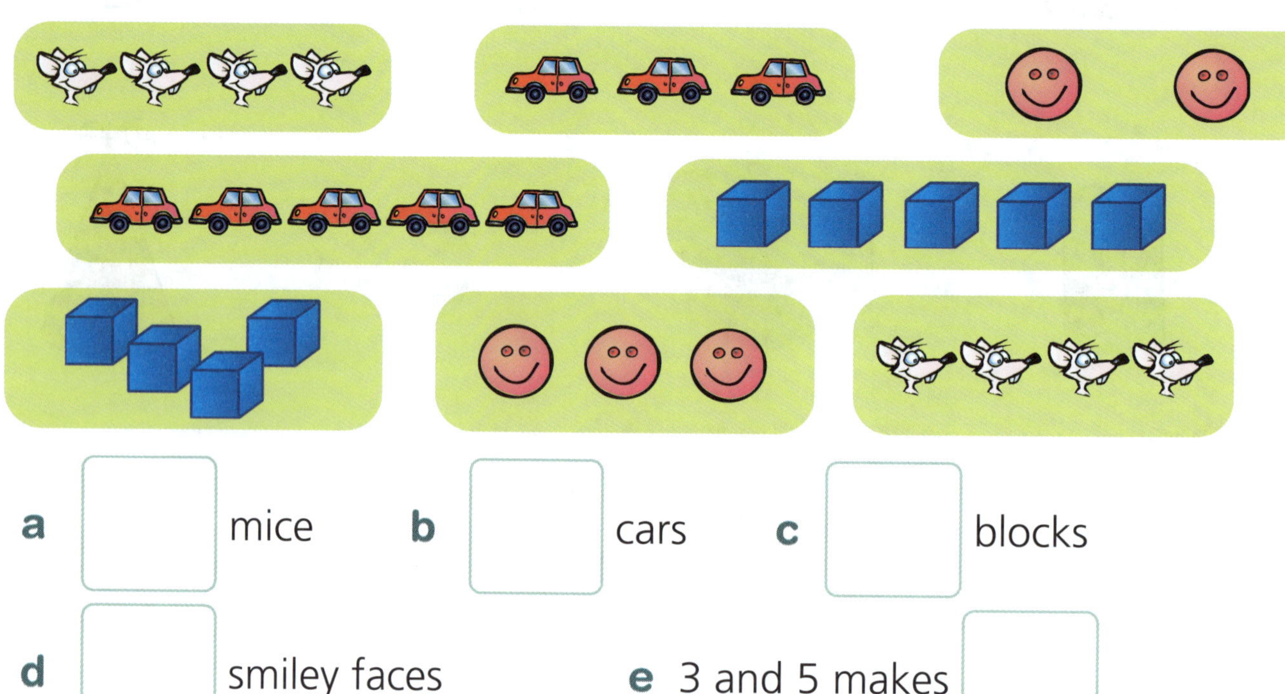

a ☐ mice   b ☐ cars   c ☐ blocks

d ☐ smiley faces   e 3 and 5 makes ☐

2 Colour and count.

2 red
4 green

☐ mushrooms

ACTIVITY

- Put some counters in box A. Count them.
- Put one counter in box B. How many altogether?
- Put another counter in box B. How many now?

Repeat this activity again and again with a different number of counters in box A.

A

B

 ISBN 9780655709015

12B Adding two groups

1 Talk about and complete these story problems.

a and 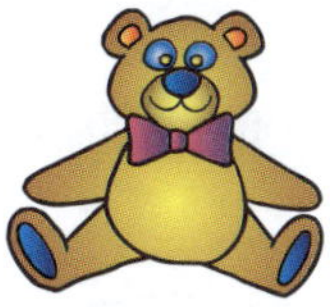 ☐ altogether

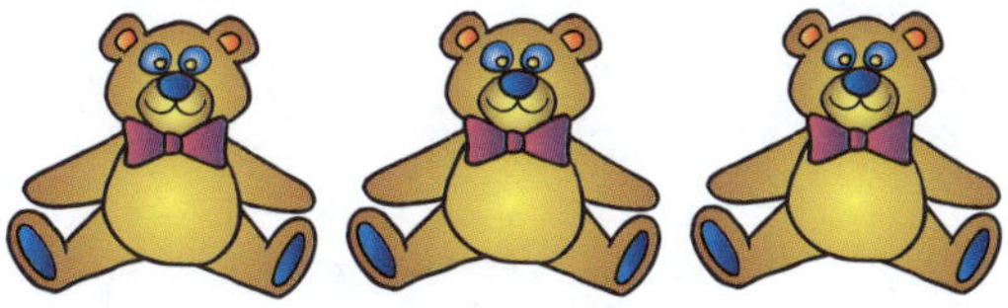

b and 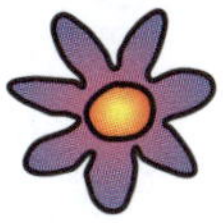 ☐ altogether

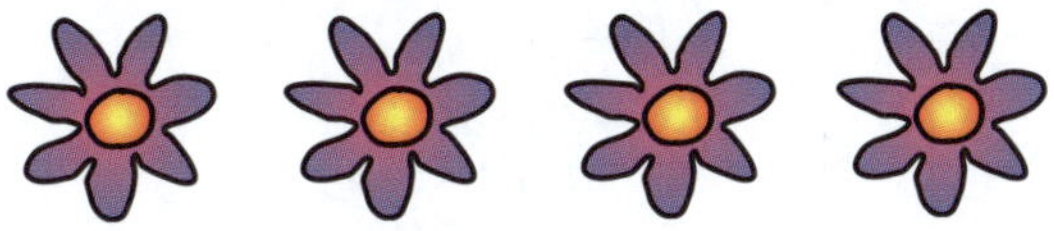

c

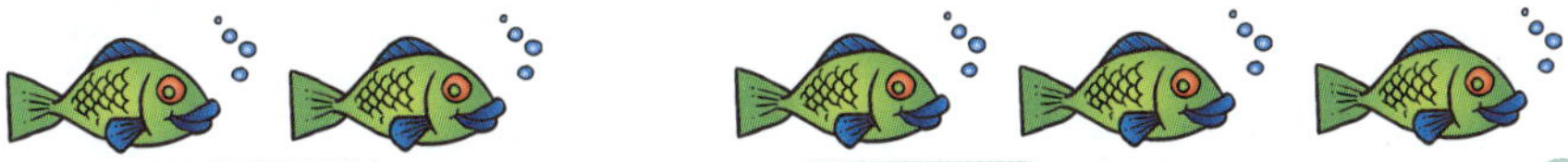

☐ fish and ☐ fish makes ☐ fish.

d

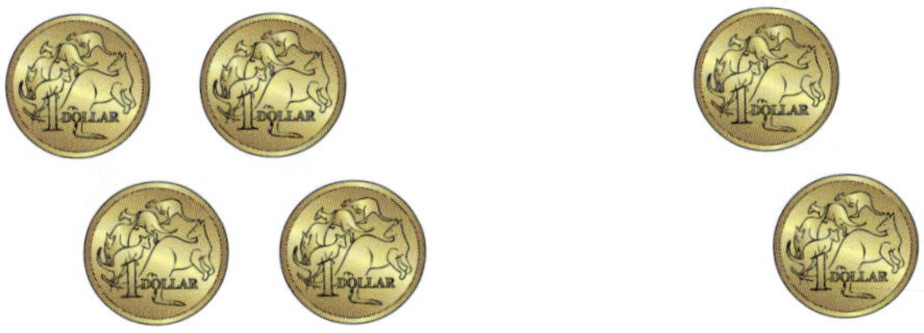

☐ coins and ☐ coins makes ☐ coins.

e

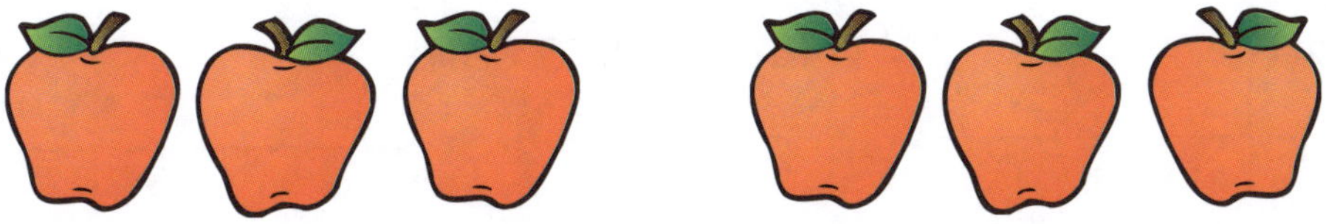

☐ apples and ☐ apples makes ☐ apples.

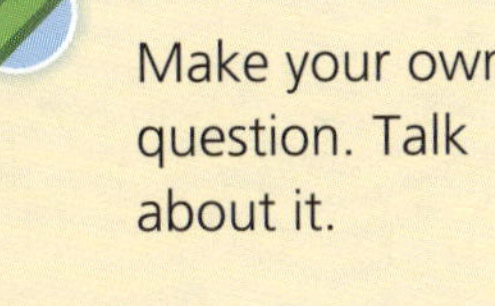

Make your own question. Talk about it.

ACTIVITY

☐ balls and ☐ balls makes ☐ balls.

12C Rolling, sliding and stacking

1. Discuss which objects you think will roll, slide or stack. Give reasons for your choice. Test your predictions using objects in question 2.

2. Colour the things that roll. **Circle** things that slide. ✔ Tick the things that stack.

Curved objects can  . Flat objects can  .

3. Draw an object that can roll. Draw an object that can slide. Draw an object that can stack.

1 Lunch Boxes

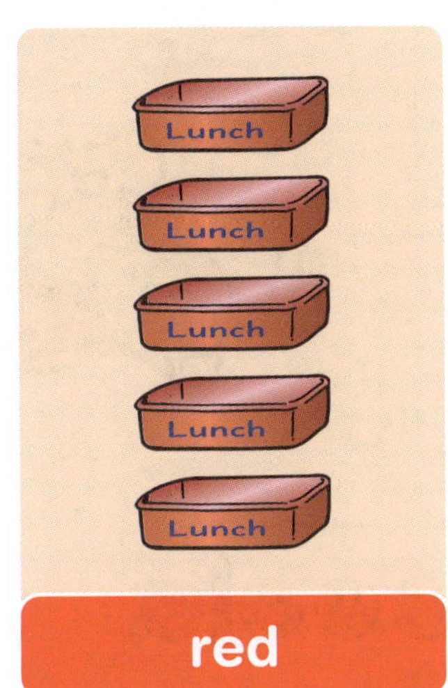

a How many red?

b How many blue?

c How many green?

d How many yellow?

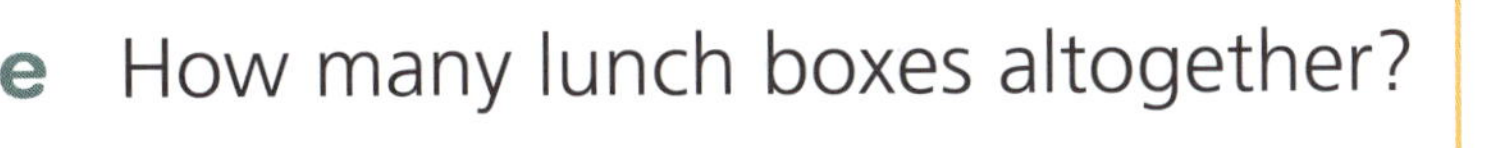

e How many lunch boxes altogether?

f Are there more red than green? **Circle** yes or no.

g Ask a friend a question about the graph.

2 Place counters or ones blocks on this display to show how many students brought these to school today, then draw faces on the display to show the results.

 • *AUSTRALIAN SIGNPOST MATHS NSW K* • ISBN 9780655709015

Using five to form numbers

1 Try to write these numbers by counting on from 5.

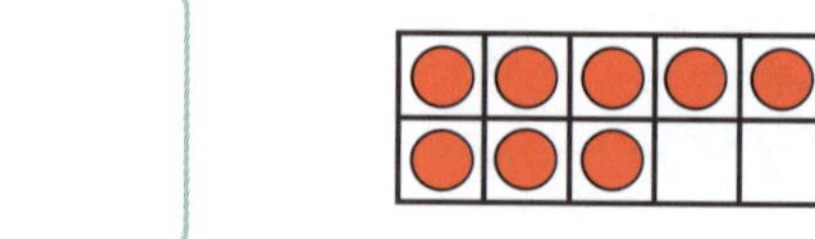
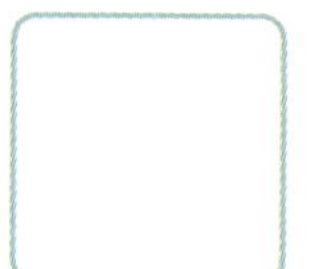

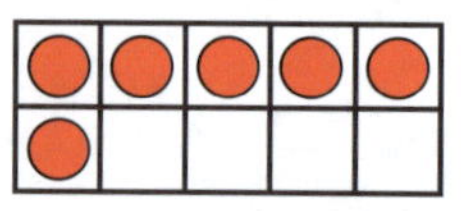

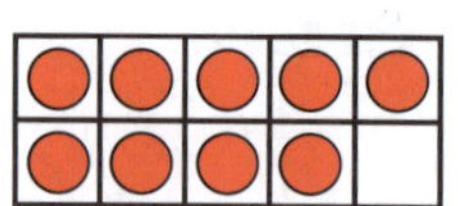

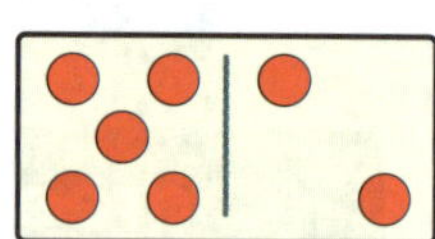

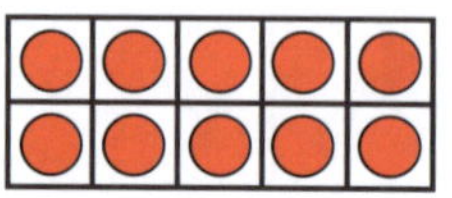

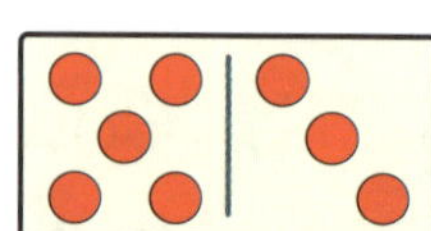

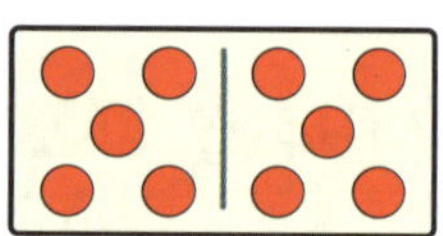

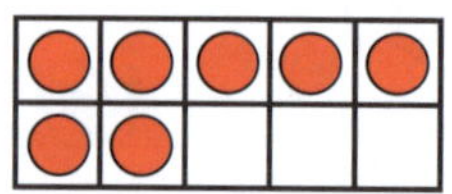

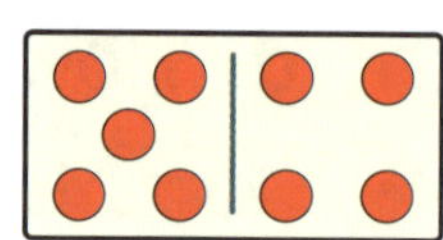

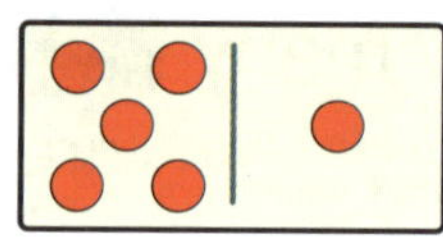

2 Use these ten frames to show the numbers.

6 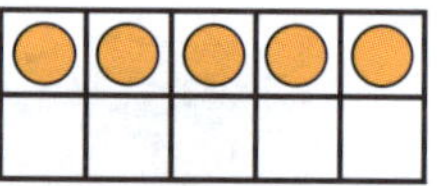

9 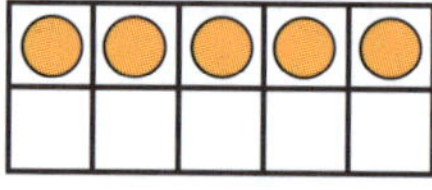

7 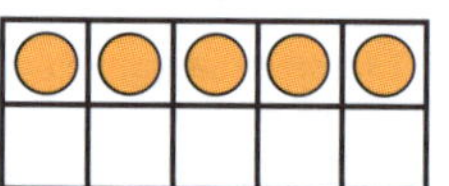

10 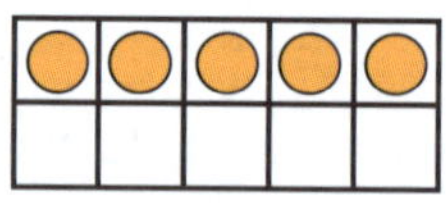

8 

5 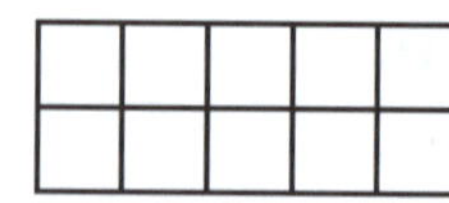

3 Use the dominoes to show the numbers.

6 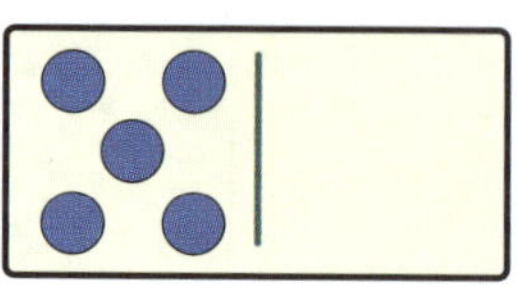

9 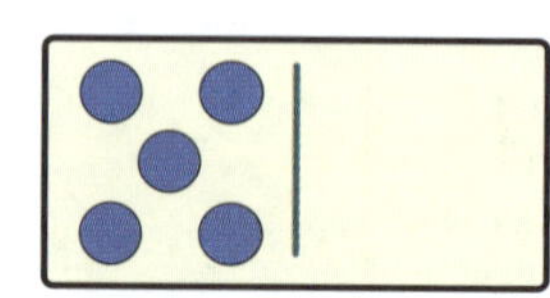

8 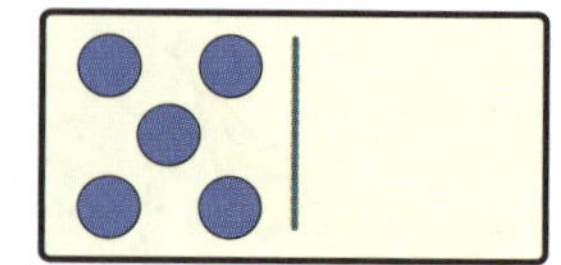

13B Adding two groups

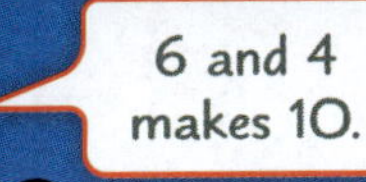

1 Complete:

a      and ☐ makes 

b      ☐ and ☐ makes 

c        ☐ and ☐ makes  

d  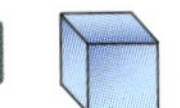     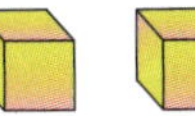 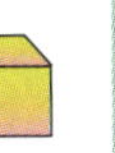 ☐ and ☐ makes  

e   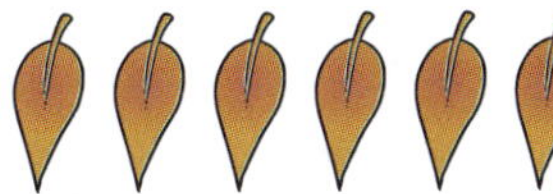 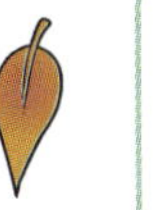 ☐ and ☐ makes 

2 Colour 2  and 5  .

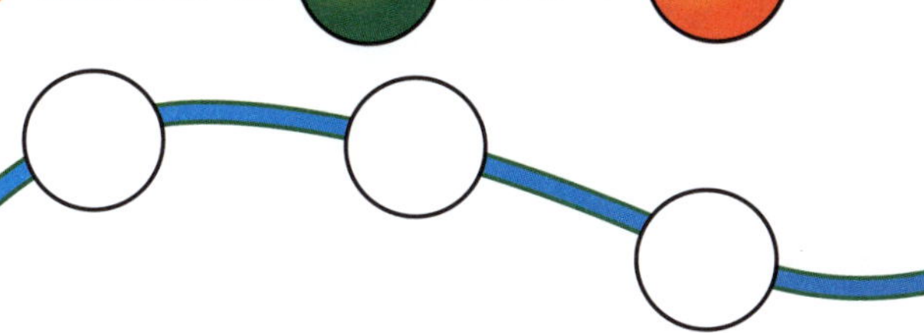

2 and 5 makes ☐

Work in groups with counters.

- Take 10 counters. Separate them into 2 groups.
- Record what you have done using a number sentence.
- Explain what you did. Do this again and again.

 • *AUSTRALIAN SIGNPOST MATHS NSW K* • ISBN 9780655709015

13C Using data displays

1

	20	
10	20	
10	20	1 DOLLAR
10	20	1 DOLLAR
10 cent coin	**20 cent coin**	**1 dollar coin**

How many 10 ?

How many 20 ?

How many 1 DOLLAR ?

Which group has the most?

Which group has the least?

How many coins altogether?

2 Use counters to make your own data display.

fish	**cars**	**balls**

13D Sequencing events in a day

1 Discuss the pictures and the time on the clocks.  Draw what you do at these times.

My Day

morning

☐ o'clock

8:00

afternoon

☐ o'clock

3:00

night

☐ o'clock

7:00

2 Put these events in order, using the numerals 1 to 3.

eating the food ☐  cooking the food ☐ washing the dishes ☐ 

14A Adding two groups

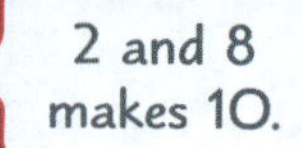

1 Complete these number sentences.

a  ☐ and ☐ makes ☐

b  ☐ and ☐ makes ☐

c  ☐ and ☐ makes ☐

d  ☐ and ☐ makes ☐

e 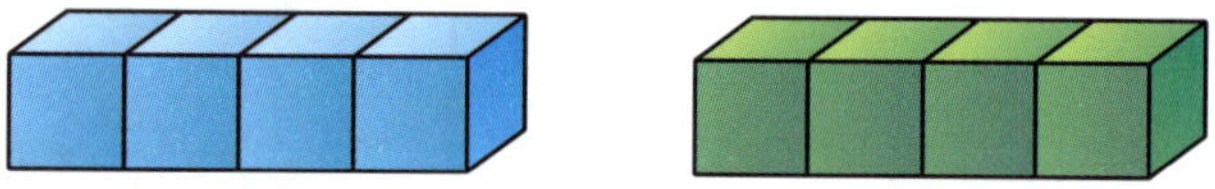 ☐ and ☐ makes ☐

ACTIVITY

Draw pictures to create your own addition stories. Complete the number sentences. Explain.

- ☐ and ☐ makes ☐
- ☐ and ☐ makes ☐
- ☐ and ☐ makes ☐

 • *AUSTRALIAN SIGNPOST MATHS NSW K* • ISBN 9780655709015

14B Adding groups

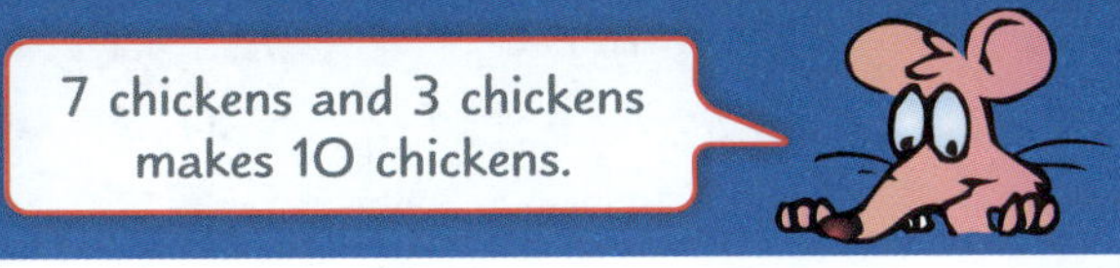

7 chickens and 3 chickens makes 10 chickens.

1 Discuss the picture.

a ☐ and ☐ makes ☐ horses altogether.

b ☐ and ☐ makes ☐ dogs altogether.

c ☐ and ☐ makes ☐ pigs altogether.

d ☐ and ☐ makes ☐ cows altogether.

e ☐ and ☐ makes ☐ trees altogether.

2 Make up your own number sentences.

14C Comparing distances

INVESTIGATION

- Colour the label to show who is closer: the captain or the pirate.
  Circle the label to show who is further away.

captain

1 2 3 6 7 5 4 9 8 10

pirate

captain

or

pirate

captain

or

pirate

captain

or

pirate

captain

or

pirate

- Join the numbers to reach the treasure. Start at 1.

 • *AUSTRALIAN SIGNPOST MATHS NSW K* • ISBN 9780655709015

14D Indirect comparison

CONCEPT

We can use string to compare lengths.

Hold the string at one end of the object and stretch it to the other end.

INVESTIGATION

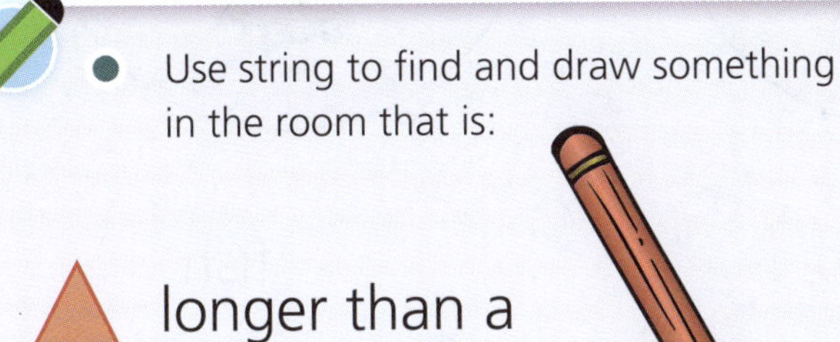

- Use string to find and draw something in the room that is:

▲ longer than a

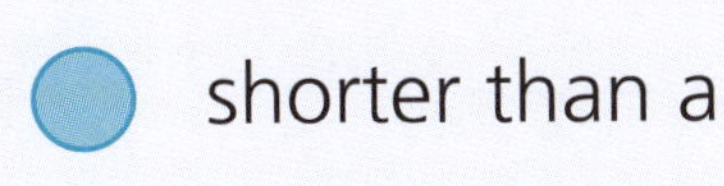

● shorter than a

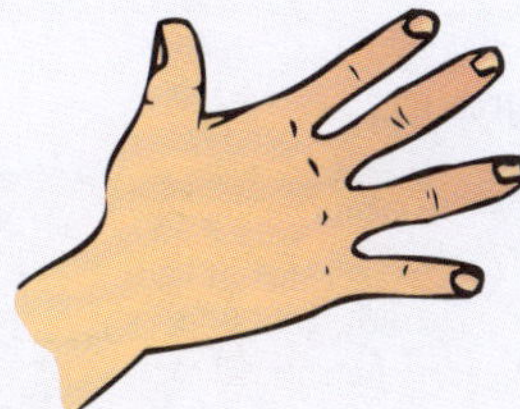

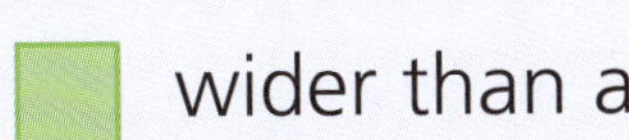

■ wider than a

◆ taller than a

- Find matching pairs of lengths in the room. Use a piece of string to compare lengths.

15A Ordering collections

1 Write the number of fish in each group. Order the groups from smallest to largest.

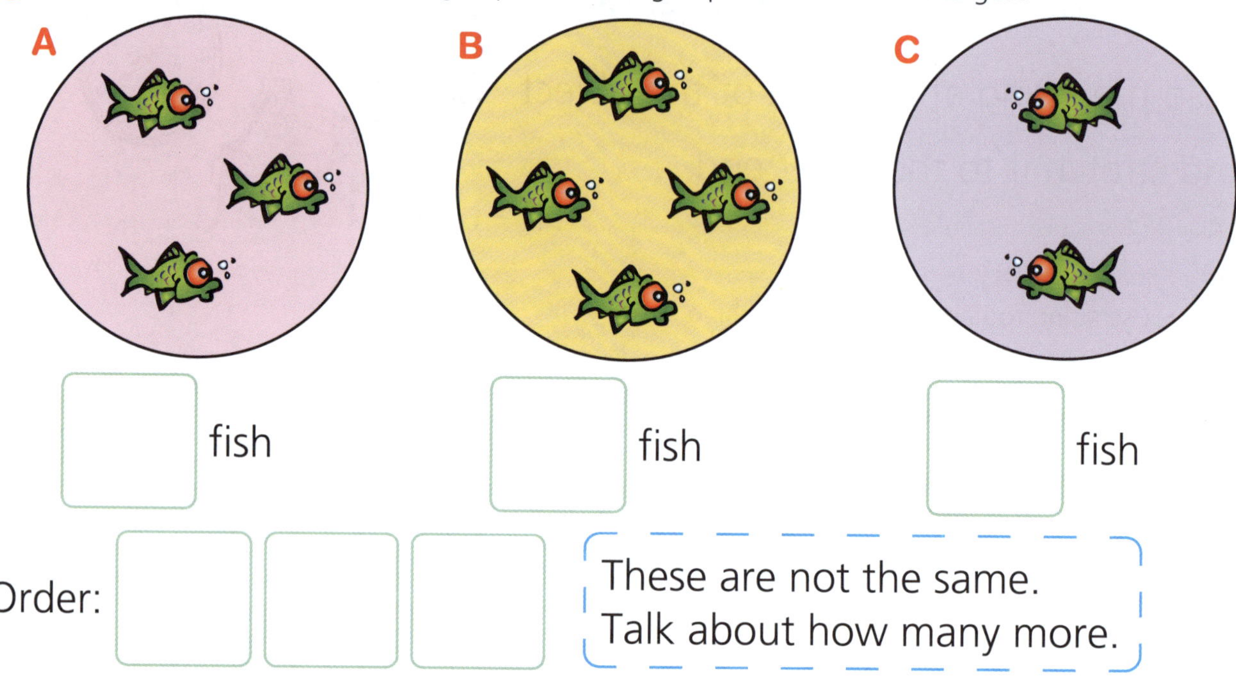

☐ fish ☐ fish ☐ fish

Order: ☐ ☐ ☐

These are not the same.
Talk about how many more.

2 Write the number of teddy bears and order the groups.

A ☐ teddy bears

B ☐ teddy bears

C ☐ teddy bears

Order: ☐ ☐ ☐

These are not the same.

Guess my number

FUN SPOT

Take turns selecting some cubes to put behind a barrier. Have your partner guess how many cubes. See how many guesses are needed to find the correct number.

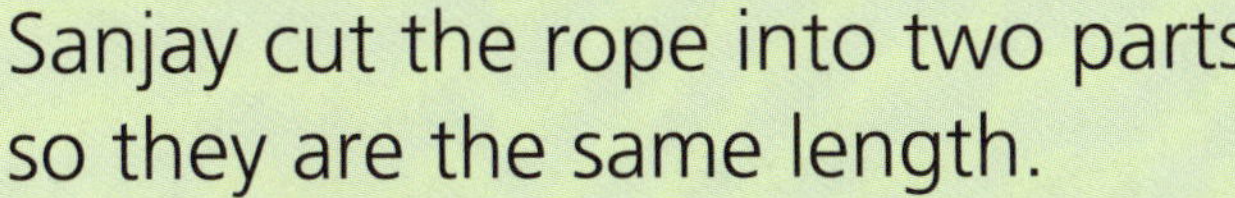

Sanjay cut the rope into two parts so they are the same length.

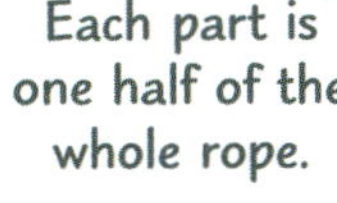

a length of rope

half a length half a length

two halves

1 **Circle** one half of each.

2 **Circle** the objects that have been cut into halves.

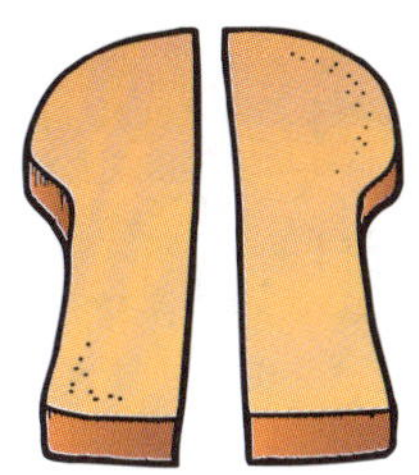

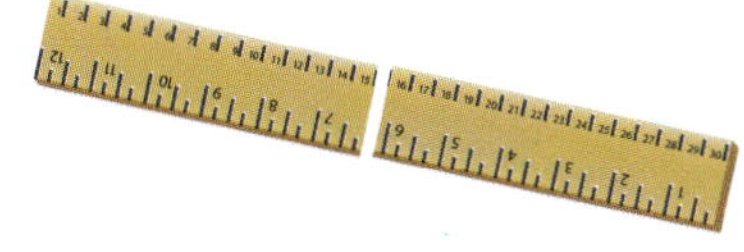

3 Draw a line to cut each object into halves.

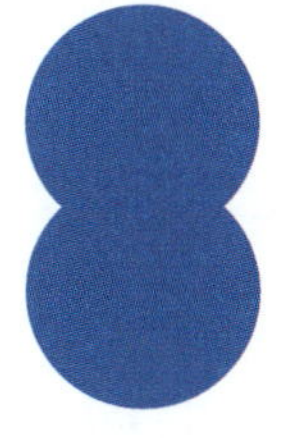

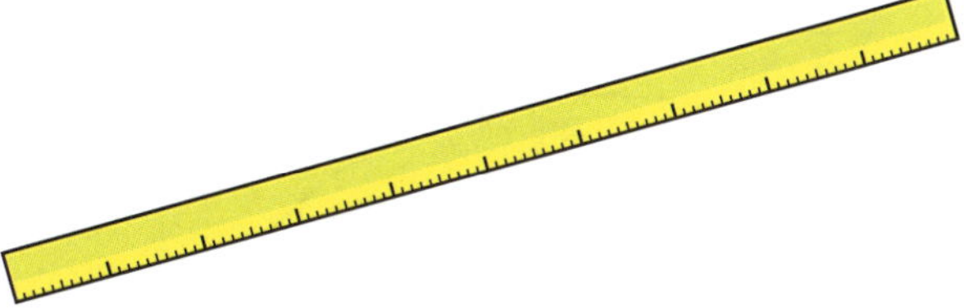

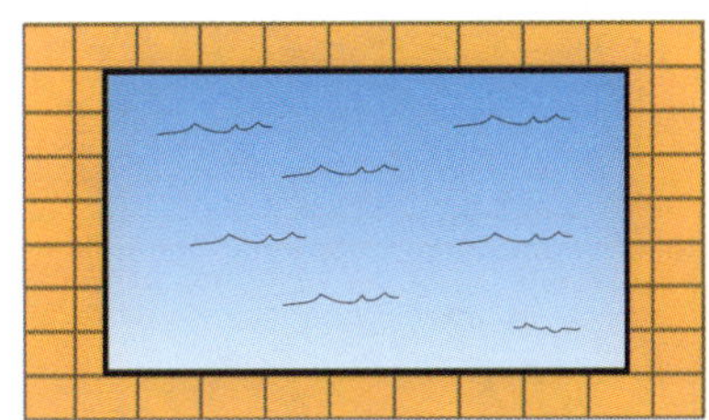

4 How many halves are in one whole?

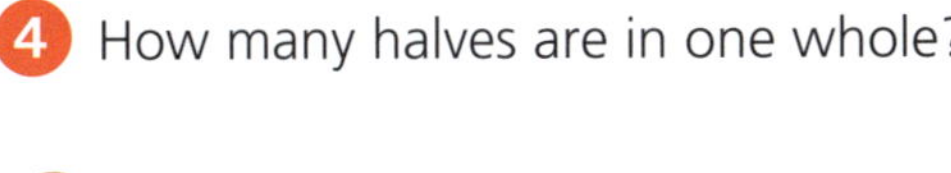

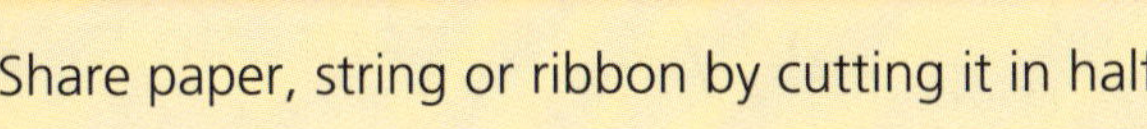

Share paper, string or ribbon by cutting it in half.

Draw what you did.

15C Clocks

1 Write the numbers.

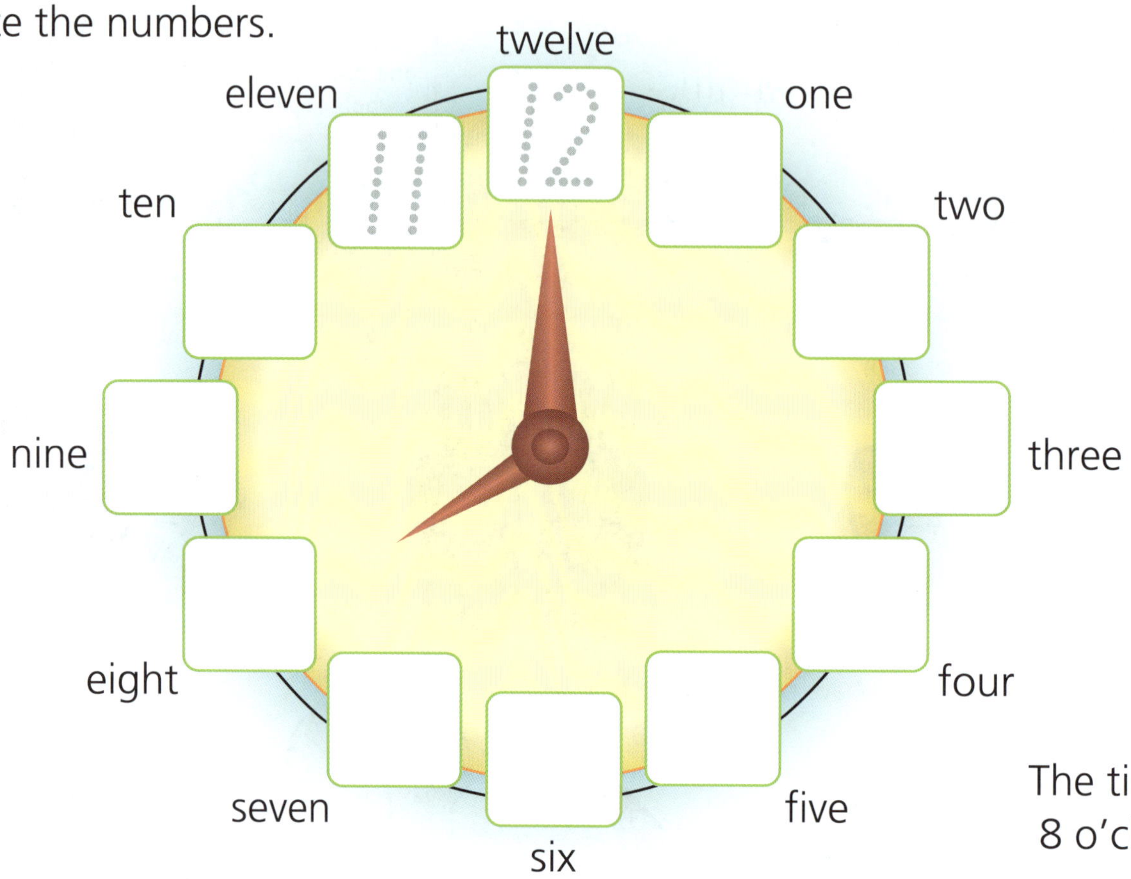

twelve 12

eleven 11

one

ten

two

nine

three

eight

four

seven

five

six

The time is 8 o'clock.

2 Write the words.

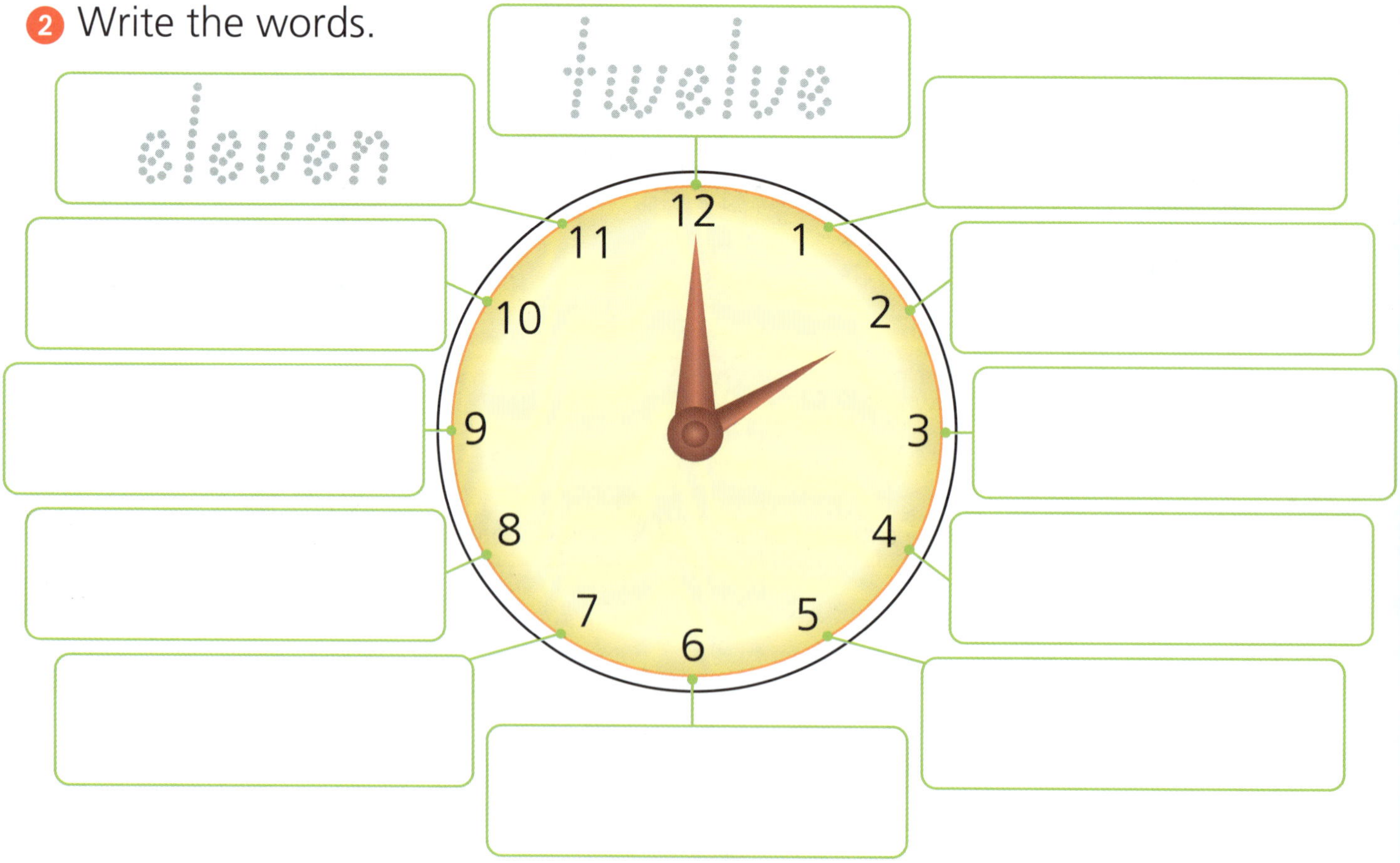

twelve

eleven

12 1 2 3 4 5 6 7 8 9 10 11

 ISBN 9780655709015

15D Telling the time

CONCEPT

This is 4 o'clock.

This also shows 4 o'clock.

1 Write the time.

☐ o'clock

☐ o'clock

☐ o'clock

☐ o'clock

☐ o'clock

☐ o'clock

2 Draw something you might do at each time.

16A Ordinal numbers

1 Complete:

first second third

2 Write the position of the kangaroo above that is:

a wearing shorts		**b** the shortest	
c carrying a bottle		**d** waving	
e wearing a hat		**f** the tallest	

3 ◯ Circle 1st place and 3rd place. ✔ Tick 2nd place and 7th place.

☐ Put a box around 4th place and 6th place.

 • *AUSTRALIAN SIGNPOST MATHS NSW K* • ISBN 9780655709015

Patterns using sounds and actions

Talk about patterns made by the sound of waves, clocks, telephones, taps and police sirens.

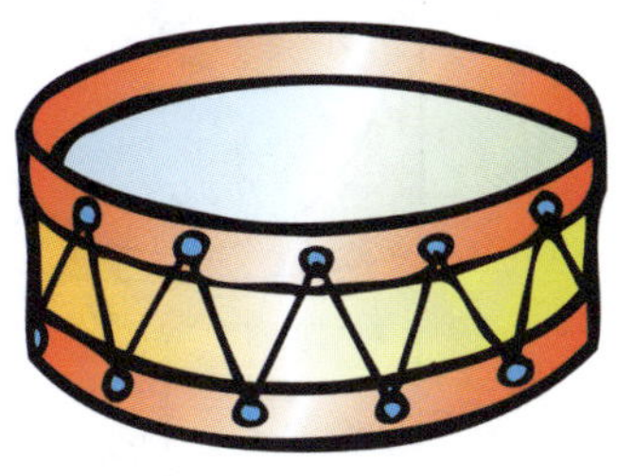

1 Talk about these sound patterns.

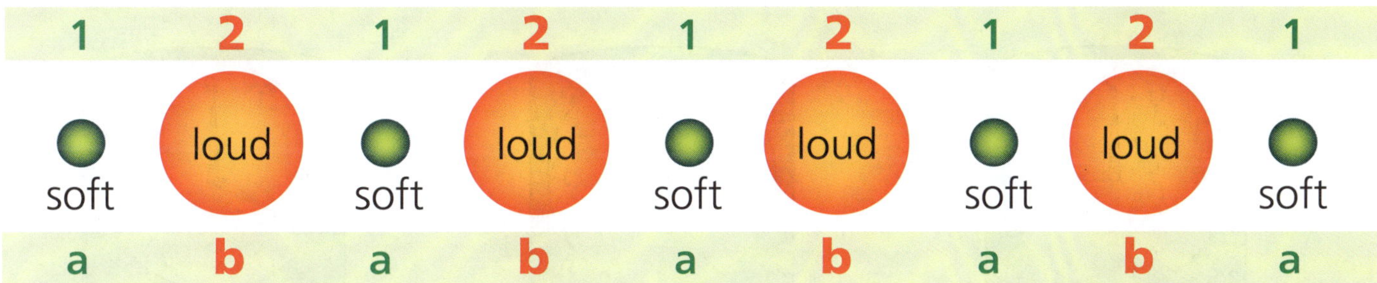

2 Say the numbers and letters to make sound patterns.

1	2	1	2	1	2	1	2	1
soft	loud	soft	loud	soft	loud	soft	loud	soft
a	b	a	b	a	b	a	b	a

Make sound patterns of your own.

3 Talk about these action patterns. Act them out with friends.

4 Draw the rest of this pattern. Act out the pattern.

Make action patterns.

16C Using o'clock

1 Write the time.

a

☐ o'clock

b

☐ o'clock

c

☐ o'clock

d

☐ o'clock

e

☐ o'clock

2 Draw hands to show the time.

a

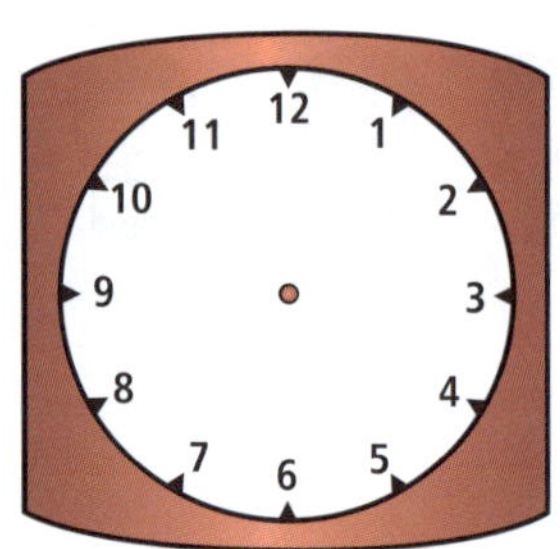

2 o'clock

b

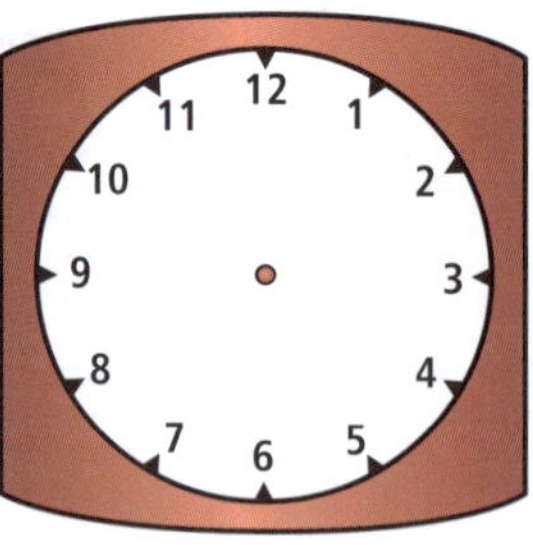

5 o'clock

c

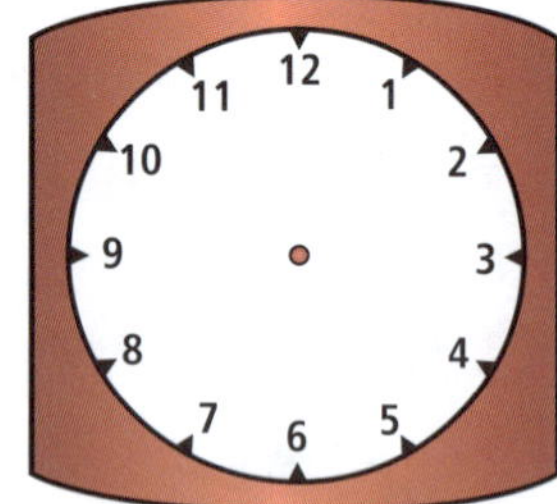

10 o'clock

d

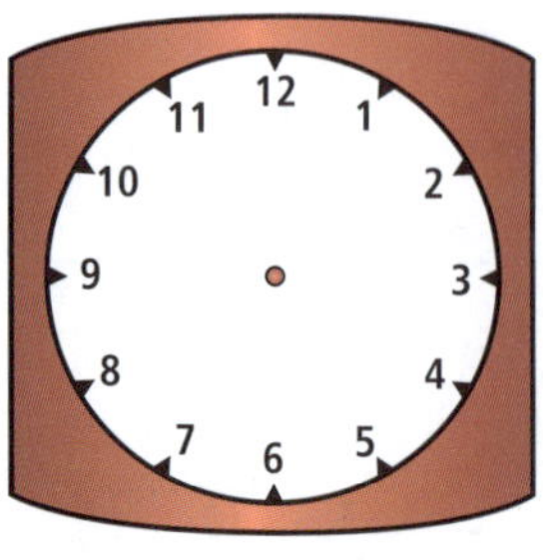

4 o'clock

e

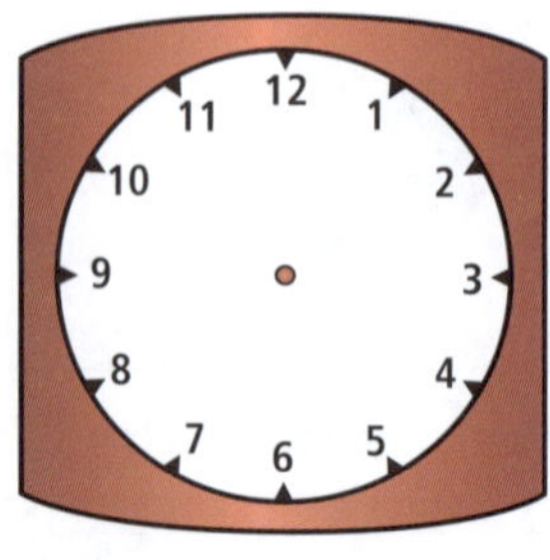

8 o'clock

f

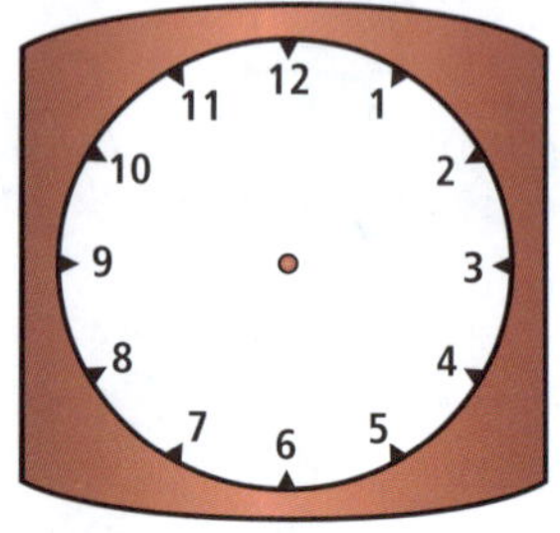

6 o'clock

 • *AUSTRALIAN SIGNPOST MATHS NSW K* • ISBN 9780655709015

16D Digital time

1 Write the time shown.

2 Draw hands to show the time.

a 3 o'clock

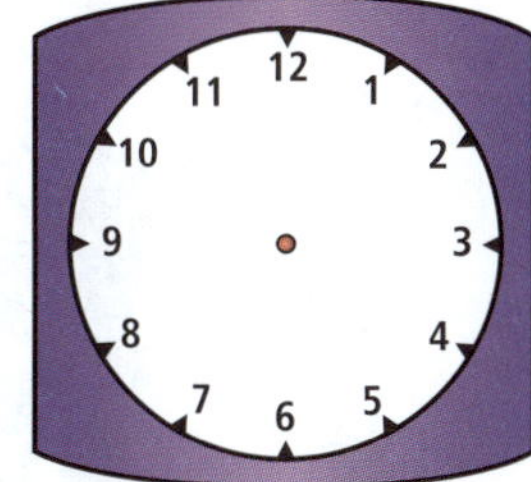

b 7 o'clock

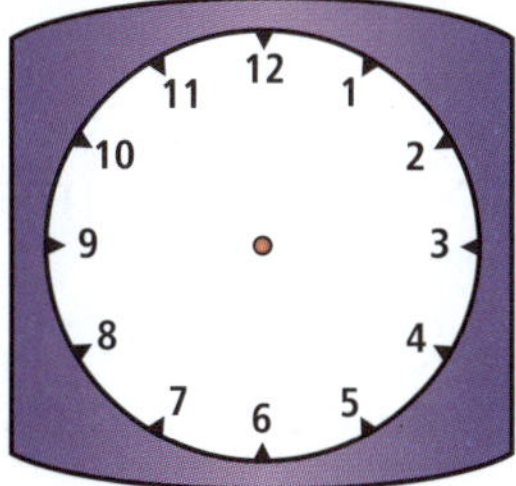

c 1 o'clock

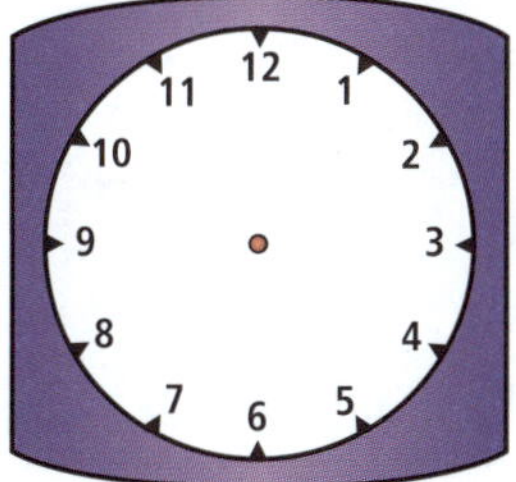

d 11 o'clock

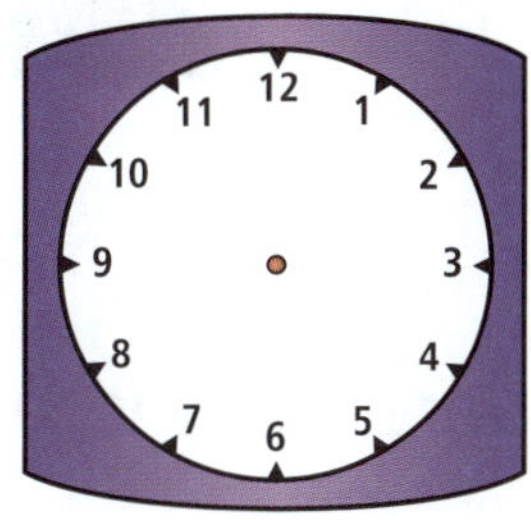

e 5 o'clock

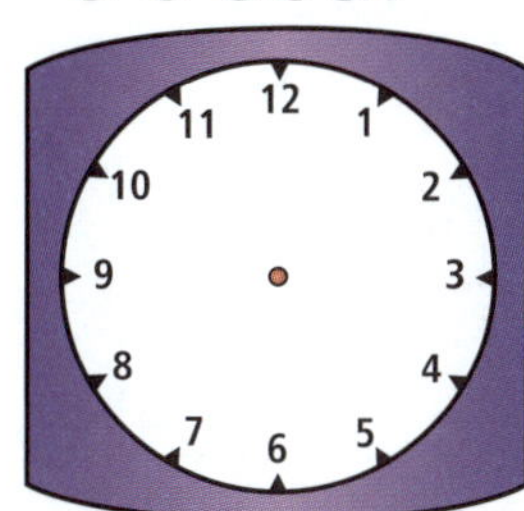

f 9 o'clock

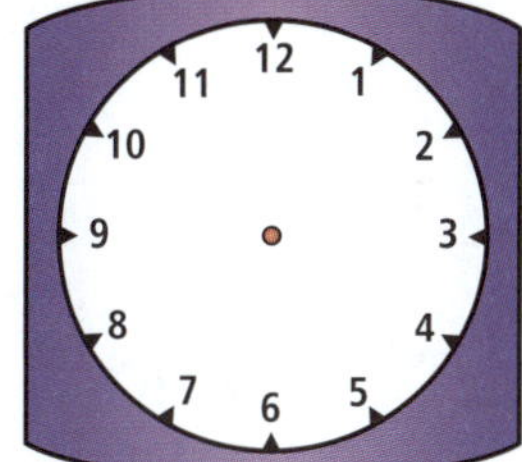

17A Adding groups

1 Count and complete.

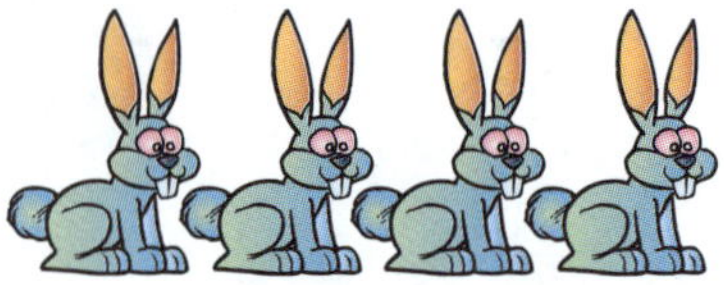

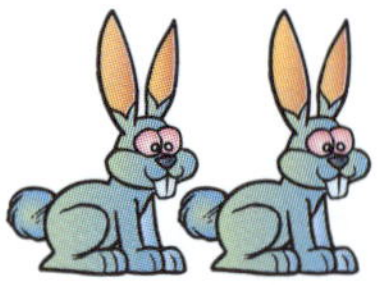

a ☐ and ☐ makes ☐

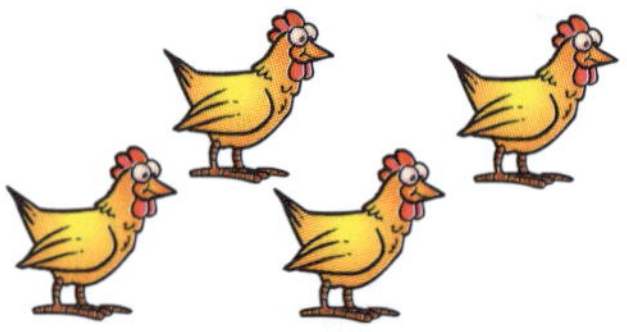

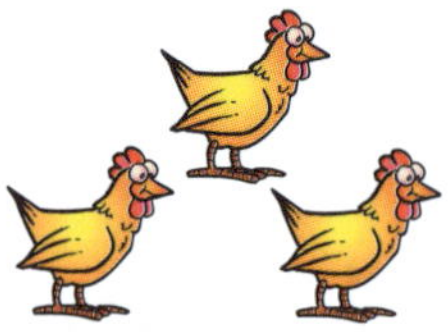

b ☐ and ☐ makes ☐

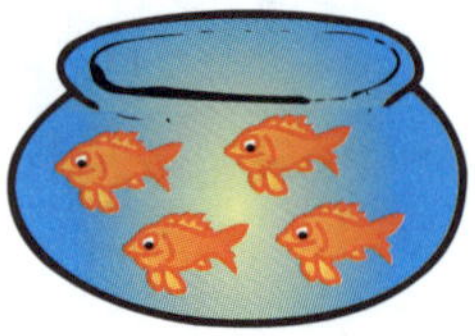

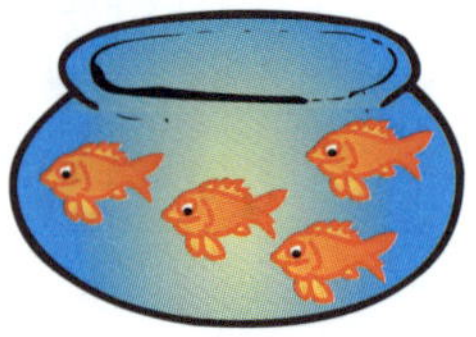

c ☐ and ☐ makes ☐

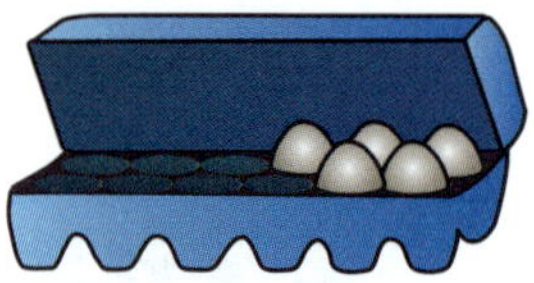

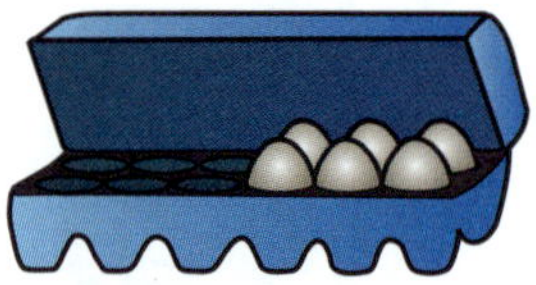

d ☐ and ☐ makes ☐

2 a Draw 2 circles for Hamid. Draw 5 circles for Mina.

☐ and ☐ makes ☐ circles altogether.

b Matt drew 3 squares. Fiona drew 4 squares. Lanh drew 2 squares.

☐ and ☐ and ☐ makes ☐ squares altogether.

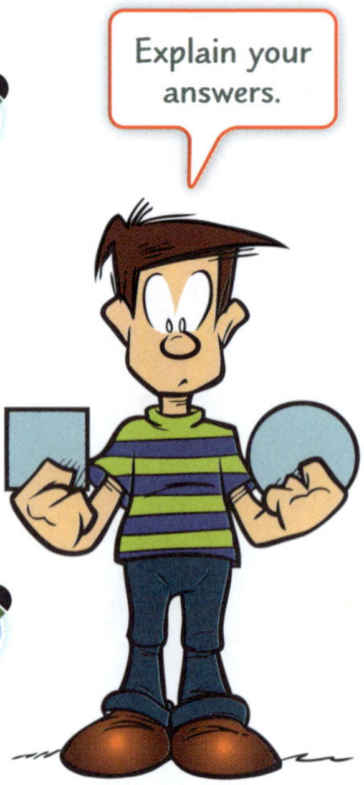

Explain your answers.

Adding groups

1 Count and complete.

a ☐ and ☐ and ☐ makes ☐

b ☐ and ☐ and ☐ makes ☐

c ☐ and ☐ and ☐ makes ☐

d ☐ and ☐ and ☐ and ☐ makes ☐

INVESTIGATION

Place 2 or 3 counters in each box. How many in each box? How many altogether?

☐ and ☐ and ☐ makes ☐

 • *AUSTRALIAN SIGNPOST MATHS NSW K* • ISBN 9780655709015

17C Ball-shaped objects

CONCEPT

This is a ball-shaped object. It is round. It has no pointy parts. It can roll.

1 Trace this ball-shaped object.

2 Circle the ball-shaped objects.

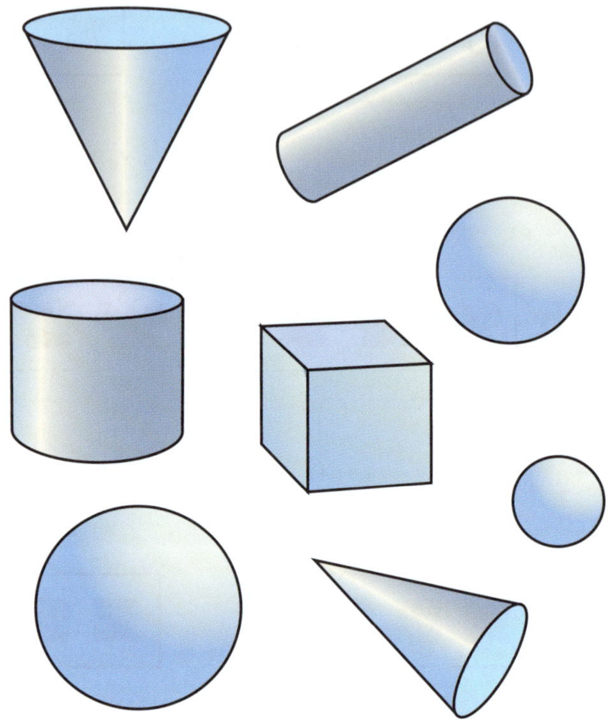

3 Circle the ball-shaped objects in this picture.

Talk about these objects using the words round, pointy, curved, straight, flat, rolls, slides and stacks.

ACTIVITY

Use playdough to make a model of a ball-shaped object. Draw your model here.

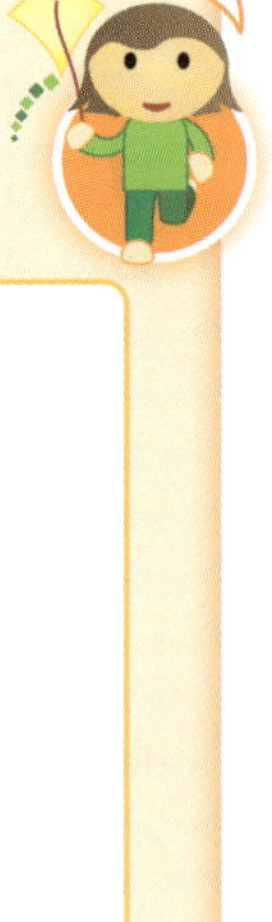

17D Sorting objects

1 Colour three objects that belong together. Discuss.

a

b

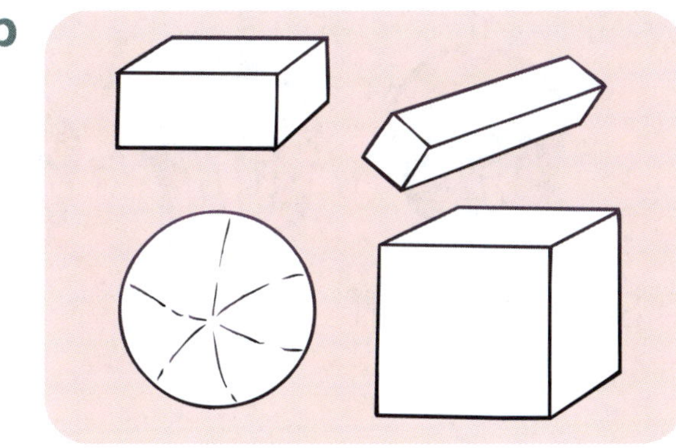

c

d

2 Circle the objects that do not belong. Discuss.

3 Colour to match objects that are the same. Circle two objects that can stack.
Give reasons for your answers.

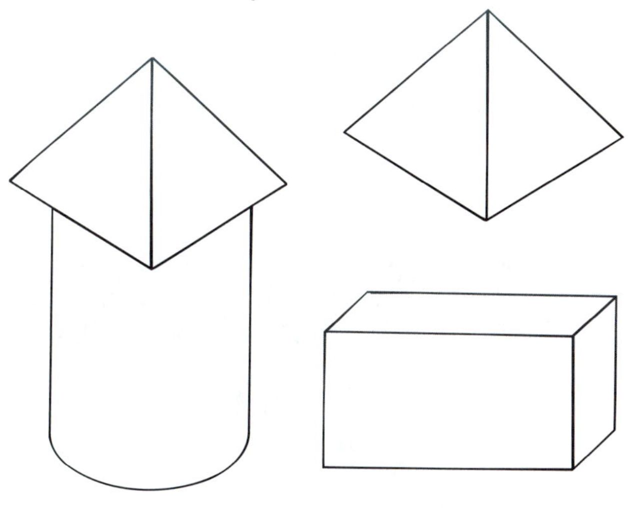

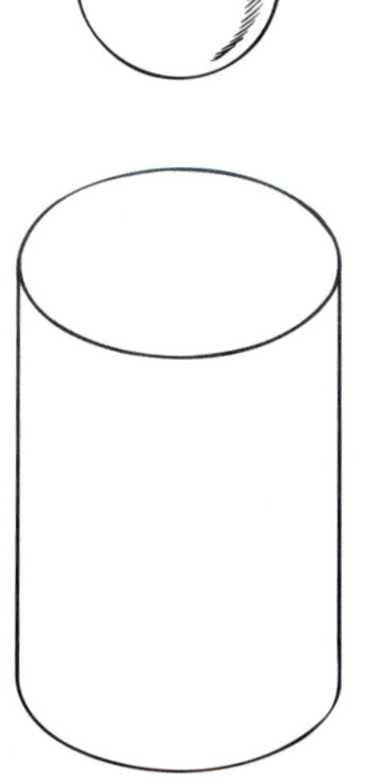

 • *AUSTRALIAN SIGNPOST MATHS NSW K* • ISBN 9780655709015

18A Taking objects away

3 take away 1 leaves 2.

CONCEPT

Jo started with 4 cupcakes. She gave 2 away.

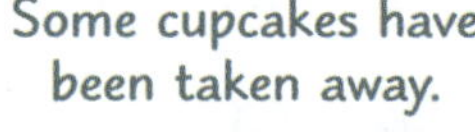

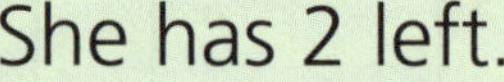

She has 2 left.

1 How many are left?

a Three dogs, one walked away.

☐ left

b Four frogs, one hopped away.

☐ left

c Six grubs, two fell off.

☐ left

d Eight birds, three flew away.

☐ left

e Ten rabbits, four hopped away.

☐ left

f Seven apples, two were eaten.

☐ left

 • *AUSTRALIAN SIGNPOST MATHS NSW K* • ISBN 9780655709015

18B Taking away

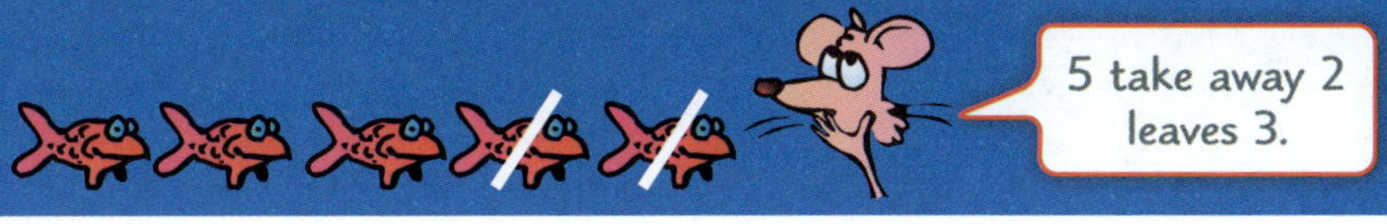

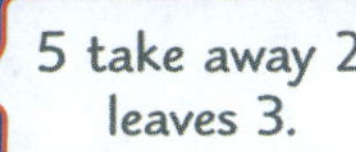

1 Complete each number sentence.

a 8 cakes take away 3 leaves ☐.

b ☐ elephants take away ☐ leaves ☐.

c ☐ ducks take away ☐ leaves ☐.

d ☐ boats take away ☐ leaves ☐.

ACTIVITY

In pairs, collect a group of 8 objects. Take turns to take some away and have your partner find how many are left. Talk about this using the number sentence below.

☐ objects take away ☐ leaves ☐.

18C Area

CONCEPT

Area is the amount of surface.

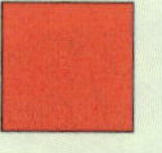

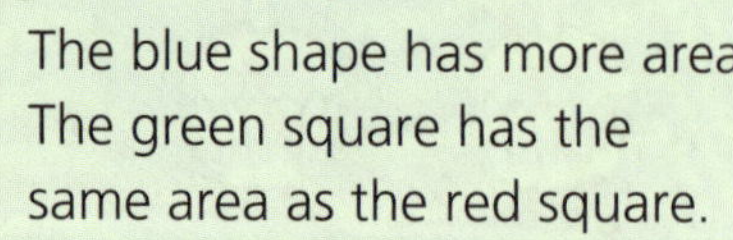

The blue shape has more area.
The green square has the same area as the red square.

The $20 takes up more space.

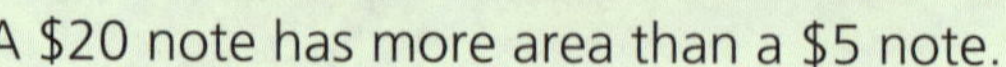

A $20 note has more area than a $5 note.

1 **Circle** the larger area in each part.

a

b

c

d

2 Colour the area of the closed shapes. Draw a star inside the biggest area.

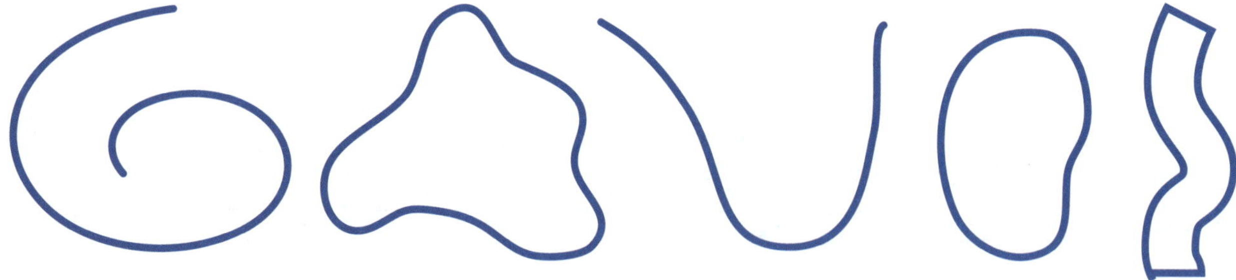

3 Draw three closed shapes. Colour the smallest area red.

Area is the inside part of a shape.

4 Cover this area with smaller shapes.

18D Days of the week

12
1
2
3
4
5
6
7
8
9
10
11

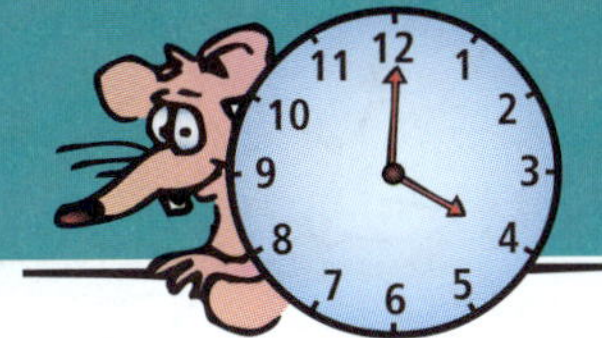

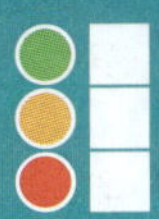

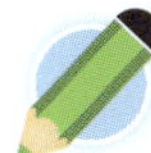

1 Help John find his way to the waterhole by following the days of the week.
Draw the route he takes. Colour the days of the weekend red and the weekdays blue.

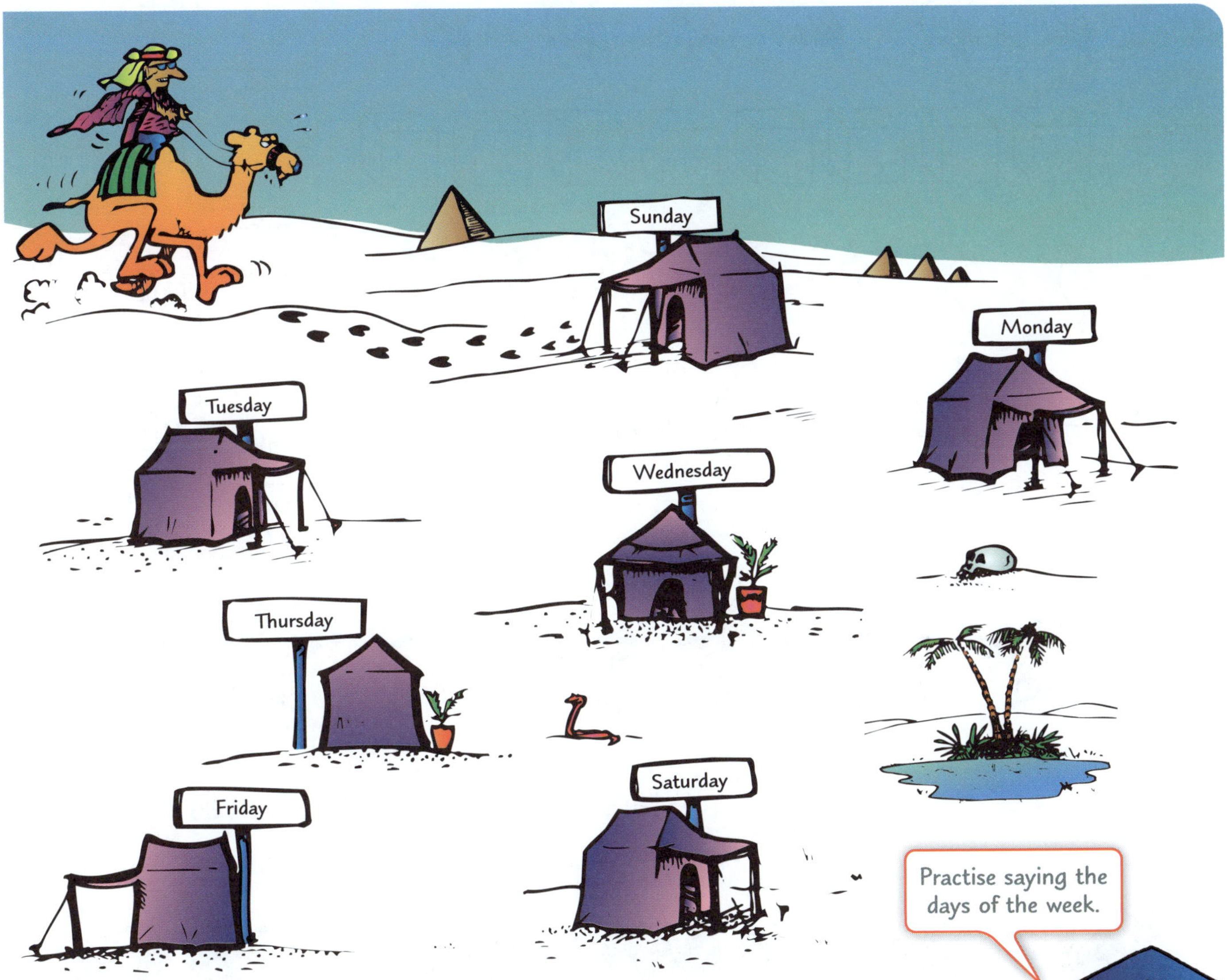
Sunday
Monday
Tuesday
Wednesday
Thursday
Friday
Saturday

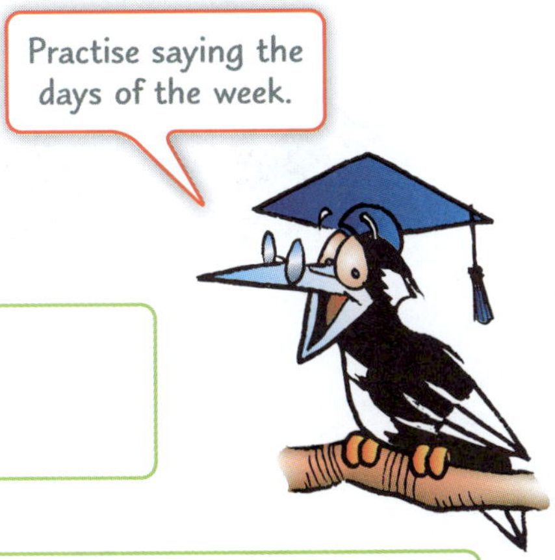
Practise saying the days of the week.

2 How many days in one week? ☐

3 How many school days in one week? ☐

4 Draw a picture of something you do on the weekend.

Saturday	Sunday

19A Taking away

1 Cross off two objects in each row. Write how many are left. Colour them in.

a [ ] left

b [ ] left

c [ ] left

d [ ] left

2 Complete each number sentence.

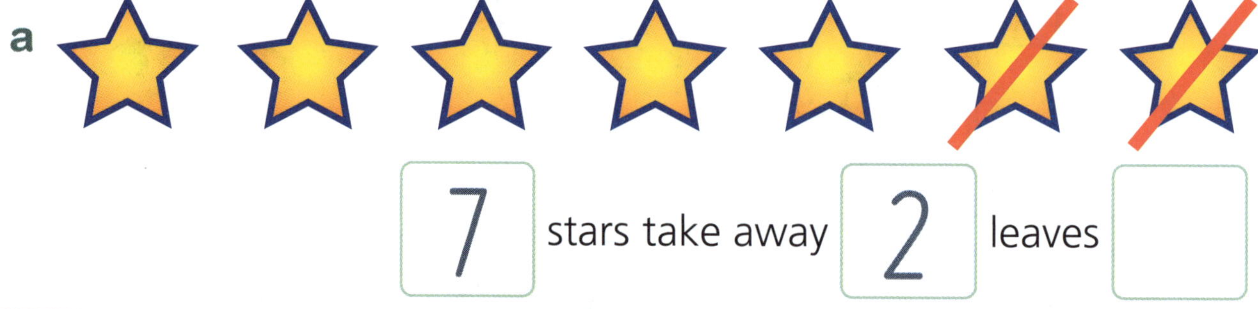

a 7 stars take away 2 leaves [ ]

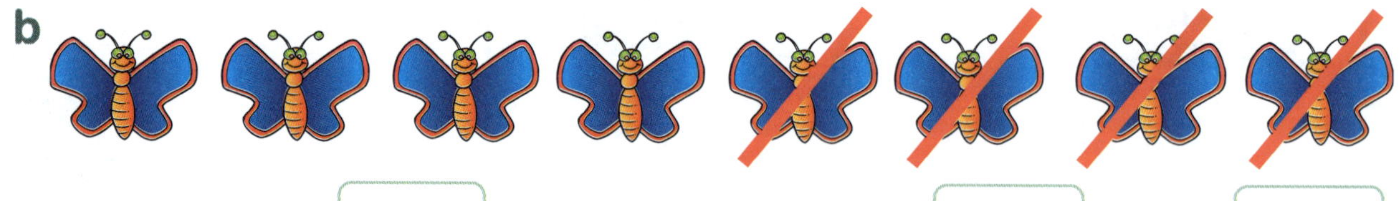

b [ ] butterflies take away [ ] leaves [ ]

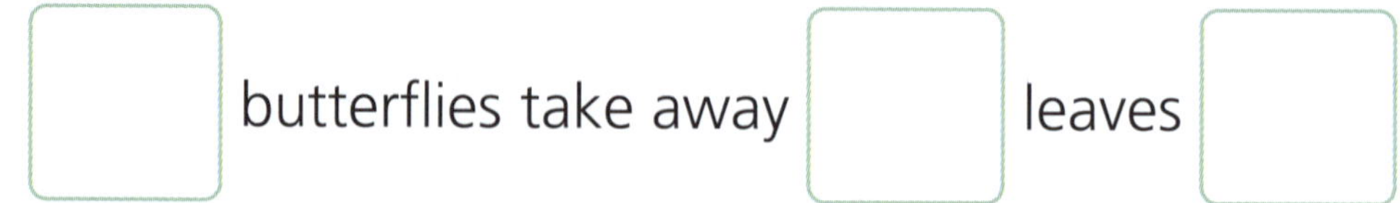

c [ ] frogs take away [ ] leaves [ ]

19B Taking away

CONCEPT

1 Complete these number sentences, then explain what you did.

a 4 take away 1 leaves ☐

b 4 take away 3 leaves ☐

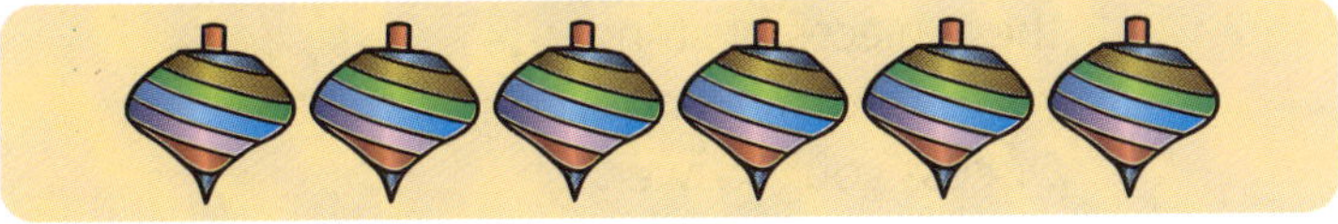

c 6 take away 4 leaves ☐

d 9 take away 2 leaves ☐

e 7 take away 4 leaves ☐

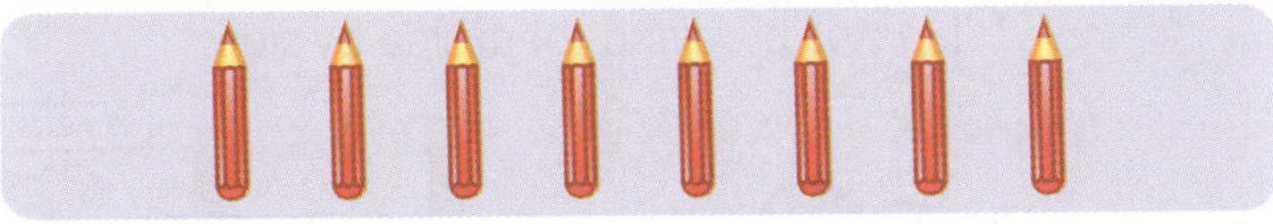

f 8 take away 3 leaves ☐

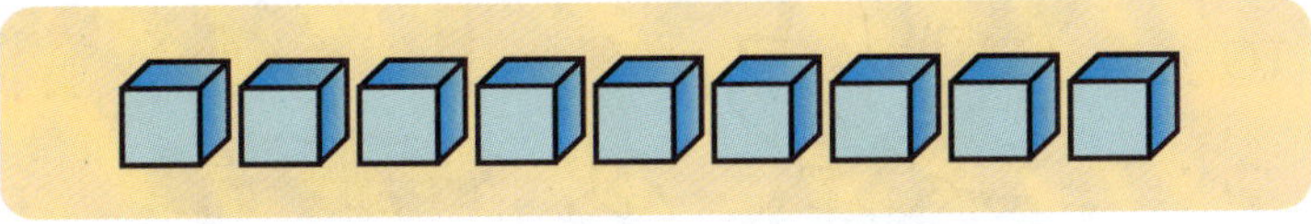

g 9 take away 0 leaves ☐

h 10 take away 8 leaves ☐

 • ISBN 9780655709015

Comparing two lengths

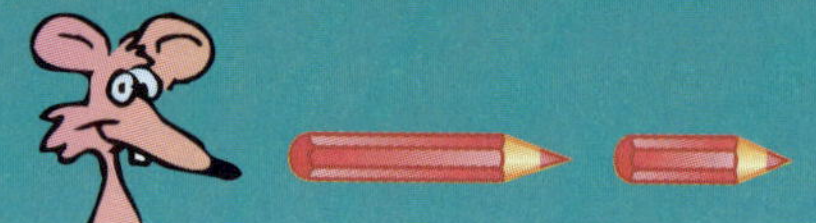

1 Circle the one that is:

deeper

thinner

lower

higher

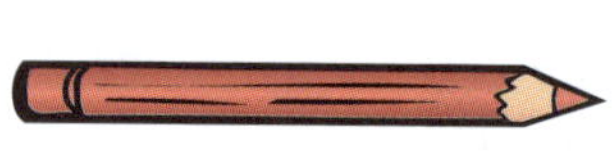

thicker

taller

2 To compare objects, place them side by side with ends matching. Colour the longer objects.

a 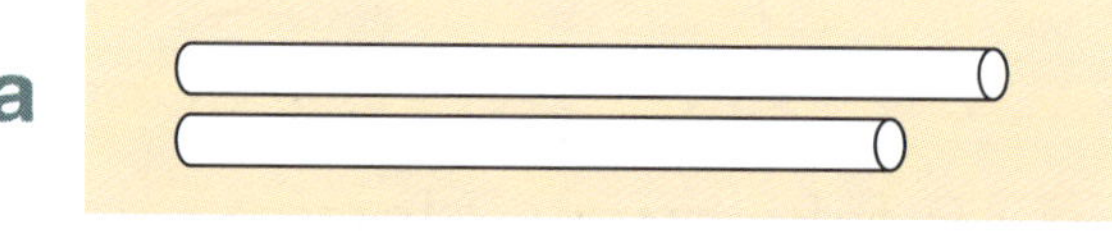

b 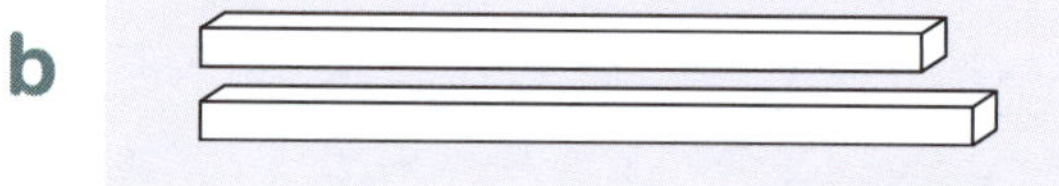

c 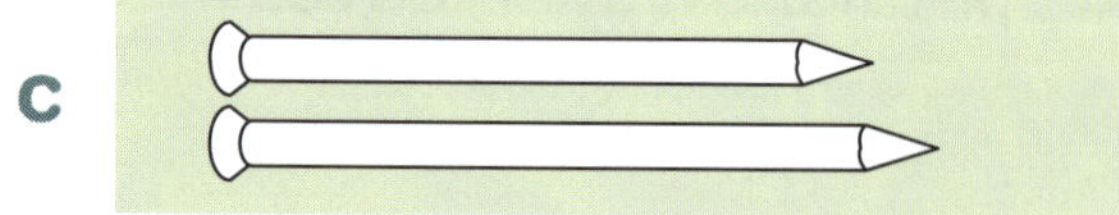

3 Do you think that these pieces of string have the same length? Discuss your answer.

4 Cut three lengths of string. Compare their lengths by laying them side by side.

Choose an object you can carry.

Compare the length of the object to the length of other objects.

Place the objects side by side and match the ends.

19D Patterns

CONCEPT

In a pattern, the parts repeat over and over again. Here the two shapes are repeated.

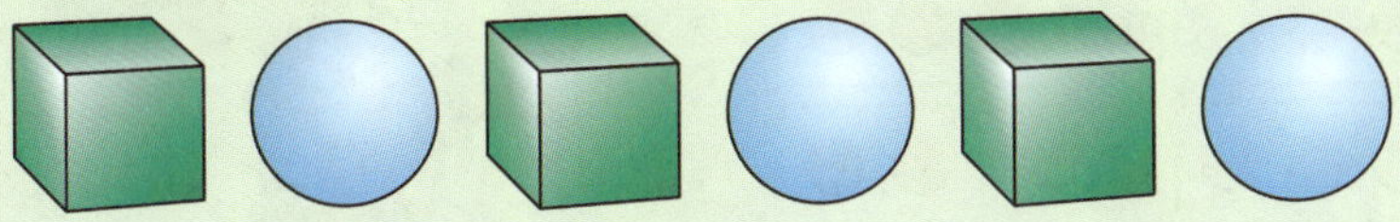

The parts could be objects, shapes, colours, sounds or actions.

1 Describe these patterns. Complete each pattern.

2 Use two colours to make a pattern. Use beads to make more patterns.

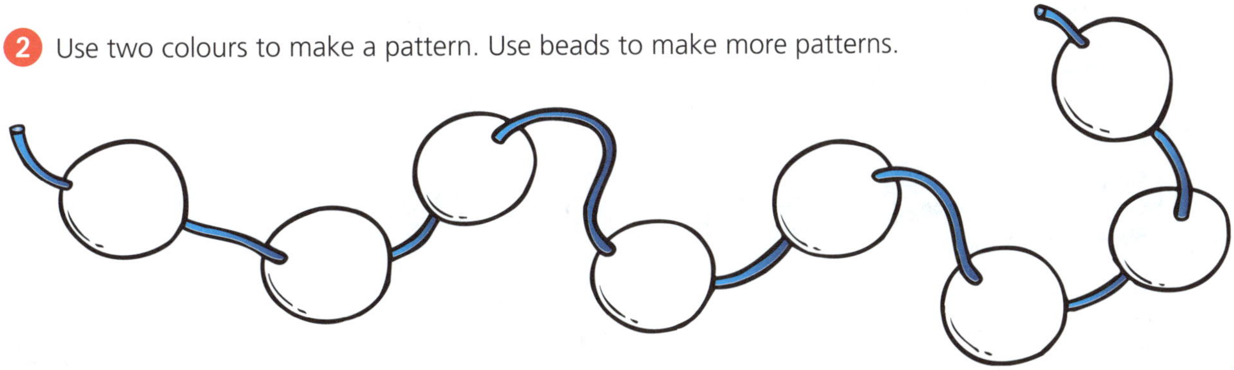

ACTIVITY

Use counters to make these patterns.
Use counters to make patterns of your own.

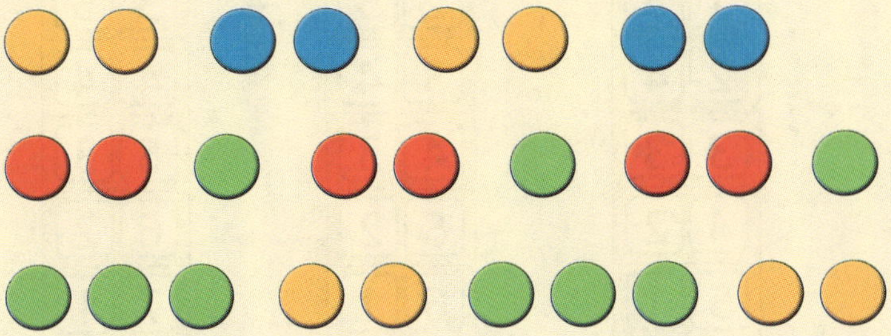

I made this one

Separating a number into parts

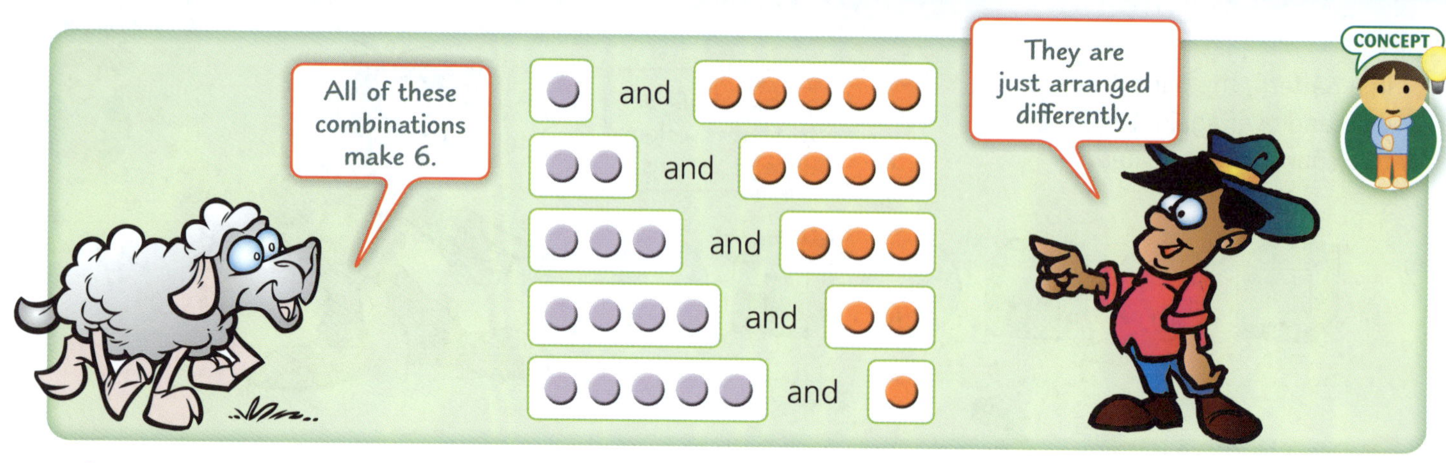

1 Discuss and complete the sentences.

 and  makes ☐

  and    makes ☐

   and   makes ☐

    and  makes ☐

     and  makes ☐

Learn what makes **6** by saying the pairs of numbers in the **6** house.

1 and 5 makes 6

2 and 4 makes 6

3 and 3 makes 6

4 and 2 makes 6

5 and 1 makes 6

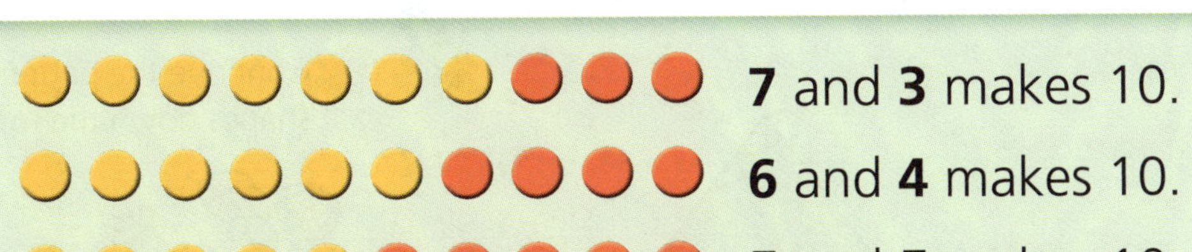

1 Complete the number sentence for each picture.

a 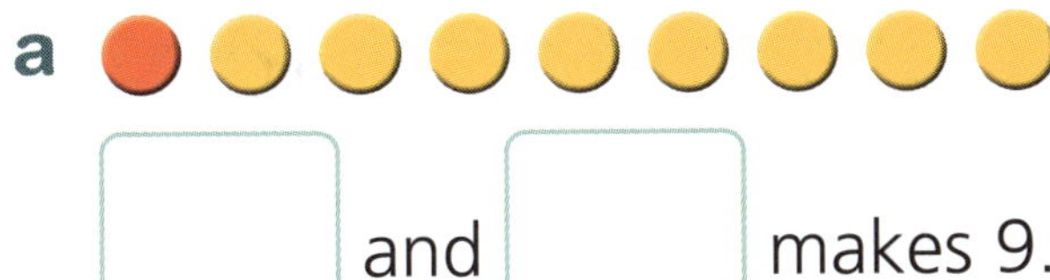

b 

c 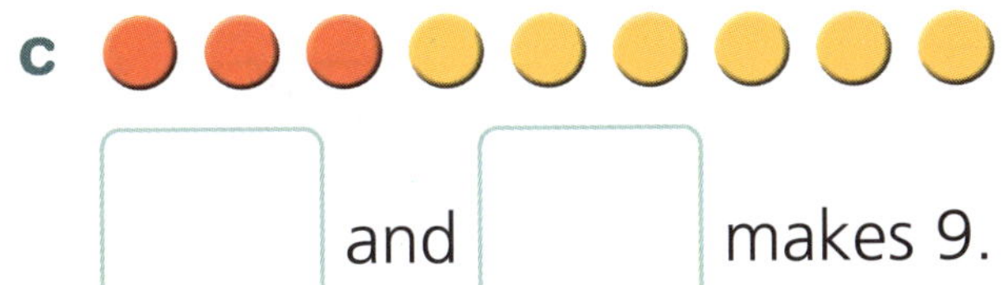

d 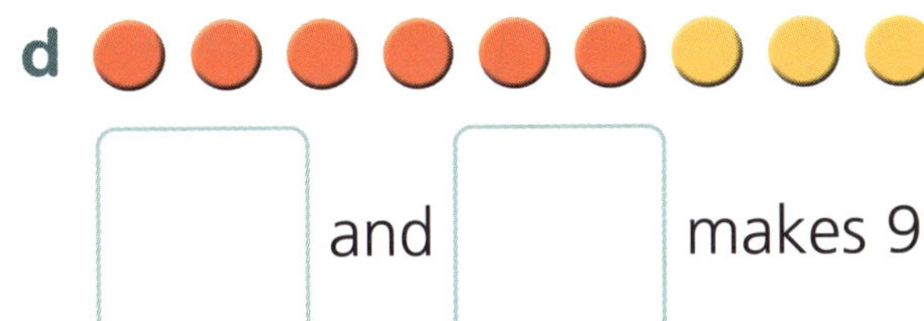

e 

f 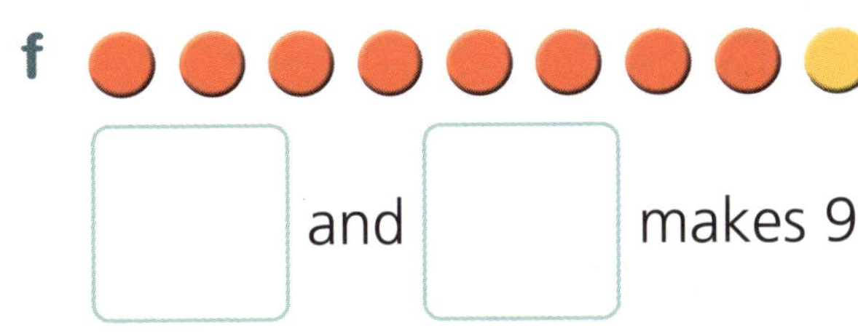

g 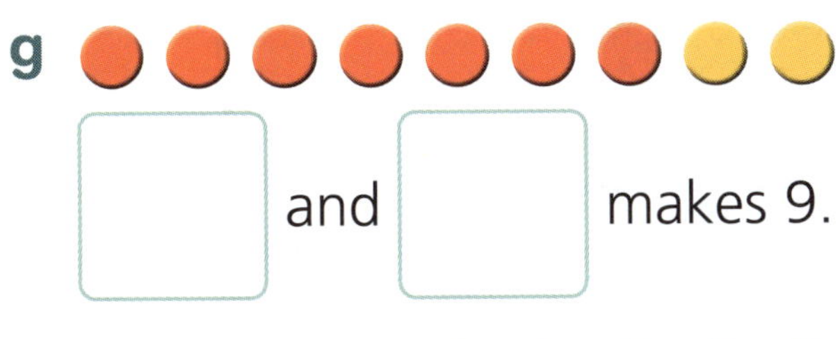

h 
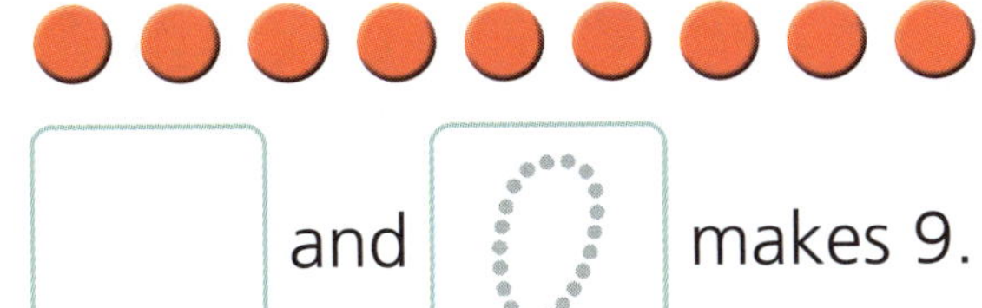

i 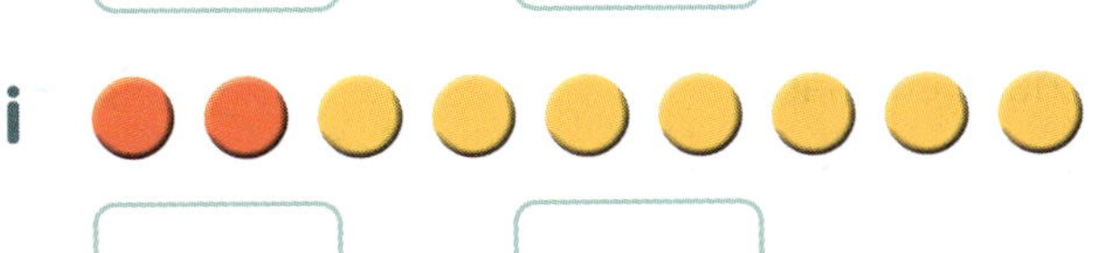
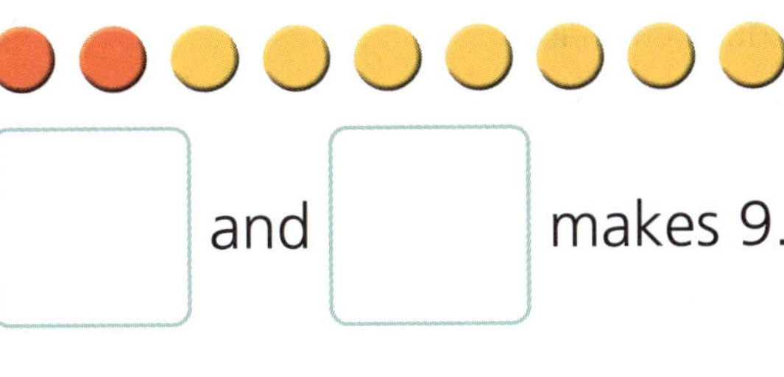

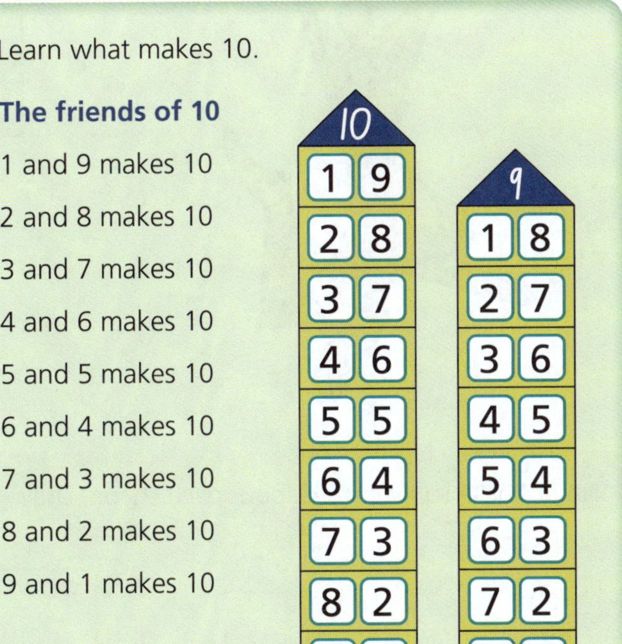

 • *AUSTRALIAN SIGNPOST MATHS NSW K* • ISBN 9780655709015

20C Shapes

1 Use colour to match shapes.

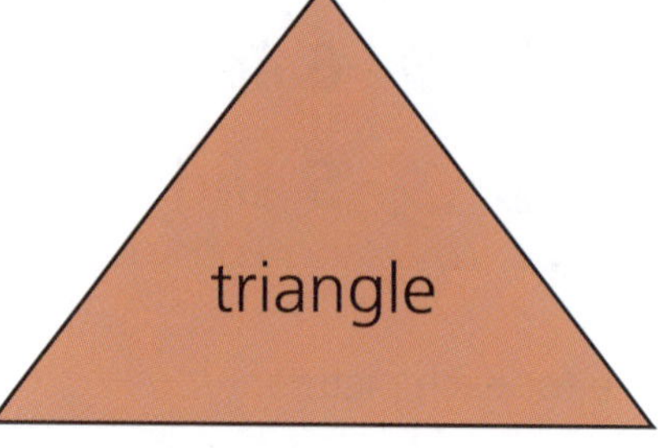

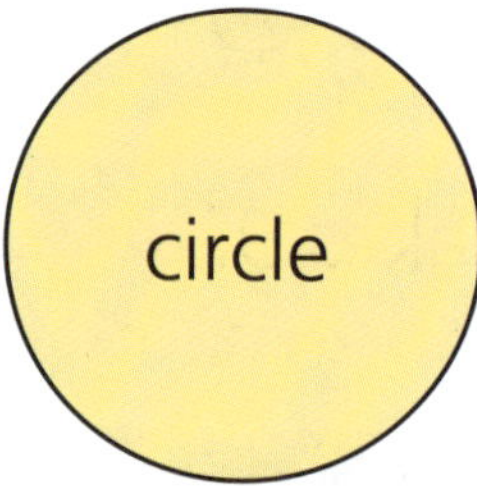

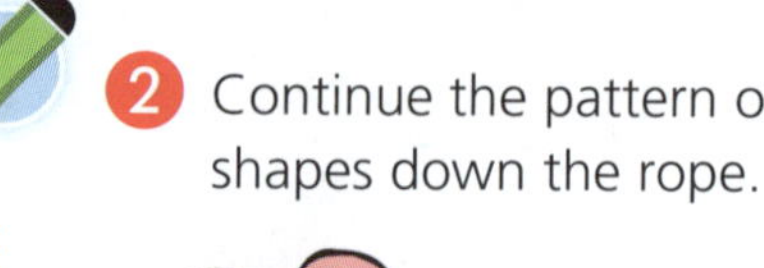

2 Continue the pattern of shapes down the rope.

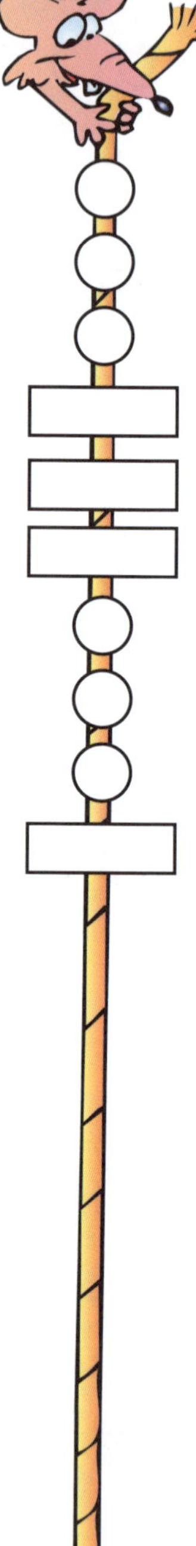

Name these shapes. Talk about each shape. Draw a circle inside each shape.

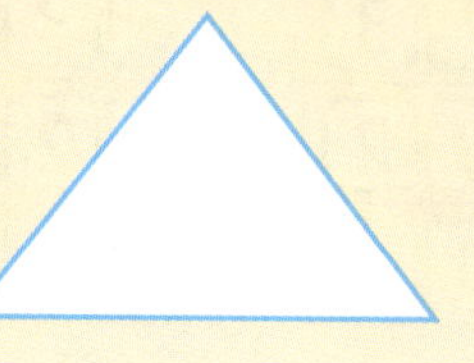

1 Party favourites

a How many liked ? ☐

b How many liked ? ☐

c How many liked ? ☐

d How many liked  ? ☐

e Colour the food liked most.

f Circle the food liked least.

g How many more students liked fruit than cakes? ☐

INVESTIGATION

Ask students to put up their hand if they think most people like Vegemite. Ten students will be chosen and asked if they like Vegemite. Put a tick next to "yes" or "no" for each student.

Do you like Vegemite? Yes or no?										
yes										
no										

 • *AUSTRALIAN SIGNPOST MATHS NSW K* • ISBN 9780655709015

21A Counting to 20

1 Discuss this picture. Count the number of flowers, chairs, hoops, windows and students.

a How many blocks? 

b How many balls? 

c How many buildings? 

d How many birds? 

e How many students?

f How many hats? 

Ask the class to start at:

- 1 and count to 20
- 20 and count backwards to 1
- 8 and count to 18
- 16 and count backwards to 6.

Ask the class:

- What number is one more than 4, 7, 11, 19?
- What number is one less than 4, 7, 11, 19?

Practise counting beyond 20.

1	2	3	4	5
6	7	8	9	10
11	12	13	14	15
16	17	18	19	20

Comparing collections

Fourteen is 1 ten and 4 ones.

Forty is 4 tens and 0 ones.

CONCEPT

Sometimes we hear patterns in the way we say numbers. This helps us remember them in order.
Clap when you hear the "teen" part of these numbers:

13, 14, 15, 16, 17, 18 and 19

1 Count each group. Write the number in each box.

Tick the largest group. **Circle** the smallest group.

A	B	C	D
* * * * * * * * * * * * * * * *	* * * * * * * * * * * * * * * * * *	* * * * * * * * * * * * * * * * *	* * * * * * * * * * * * * * * * * * *

2 Discuss the groups above. Write them in order, from smallest to largest.

INVESTIGATION

Draw your own set of different groups.
Tick the largest group. **Circle** the smallest group.

21C Cone-shaped objects

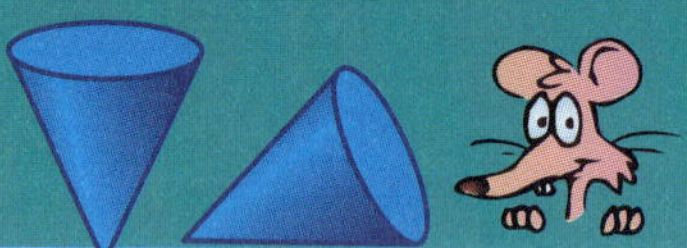

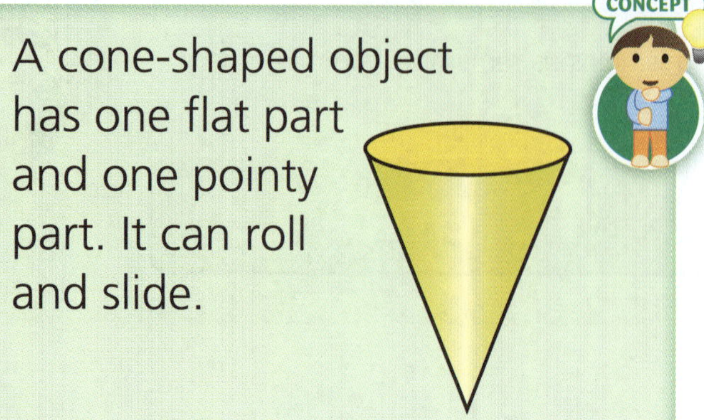

1 Trace this cone-shaped object.

3 **Circle** the cone-shaped objects in this picture.

Talk about these objects using the words round, pointy, curved, straight, flat, rolls, slides and stacks.

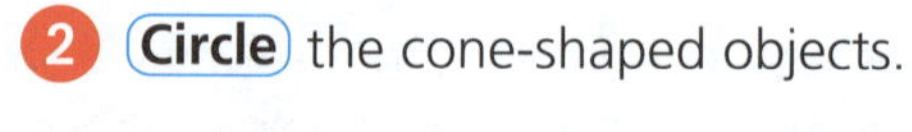

2 **Circle** the cone-shaped objects.

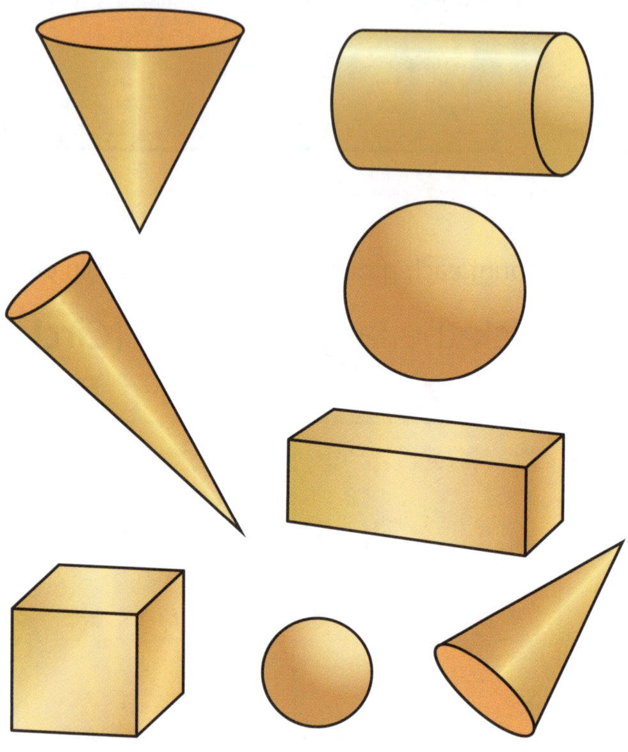

Use playdough to make a model of a cone-shaped object.
Draw your model here.

21D Left and right

The cat is on the **left**.

The dog is on the **right**.

We can make an L with our left hand.

1. Talk about objects that are on your left-hand side and those on your right.
2. Talk about the picture below using the words "left" and "right".

3. Colour the boat on the the left. Colour the cloud on the right.
   Colour the island on the left. Colour the bucket on the right.

4. Circle the left hands.

5. **Circle** your answer. "I hold my pencil in my **right** / **left** hand."

22A Groups of equal size

1 Colour the group with the most objects. Talk about 'how many more'.

a

b

c

2 Make the groups the same. Talk about equal groups.

a

b

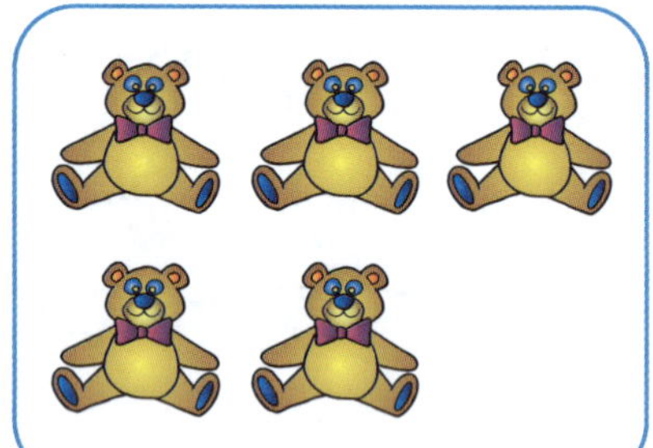

Handful of blocks

Each student takes a handful of blocks. They place their blocks in rows, side by side. Change the number of blocks to make the two groups equal. Explain your answer.

22B Matching equal groups

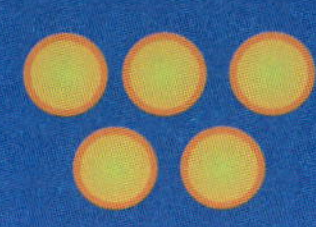

1 Match these groups so that the number on one side is the same as the number on the other side.

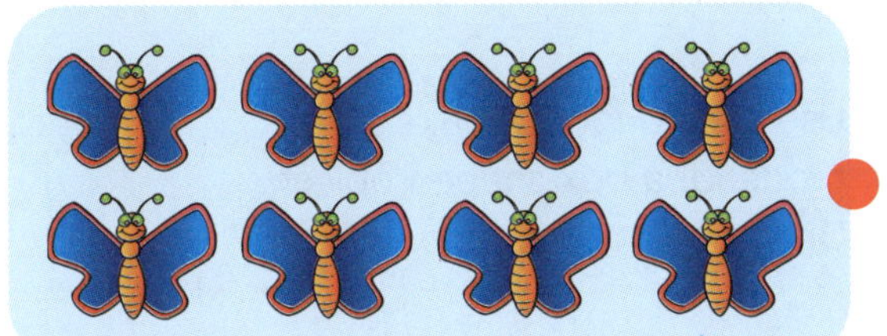

Making equal groups

Place 8 counters in a row. Under this row make another row of 8 counters. Make other equal rows of counters to show equal groups.

22C Box-shaped objects

CONCEPT

This is a box-shaped object. It has flat parts and sharp corners. It has straight edges and it can slide.

1 Trace.

If all faces are squares, it is a cube.

2 Circle the box-shaped objects.

3 Circle the box-shaped objects in this picture.

Talk about these objects using the words round, pointy, curved, straight, flat, rolls, slides and stacks.

ACTIVITY

Use playdough to make a model of a box-shaped object. Draw your model here.

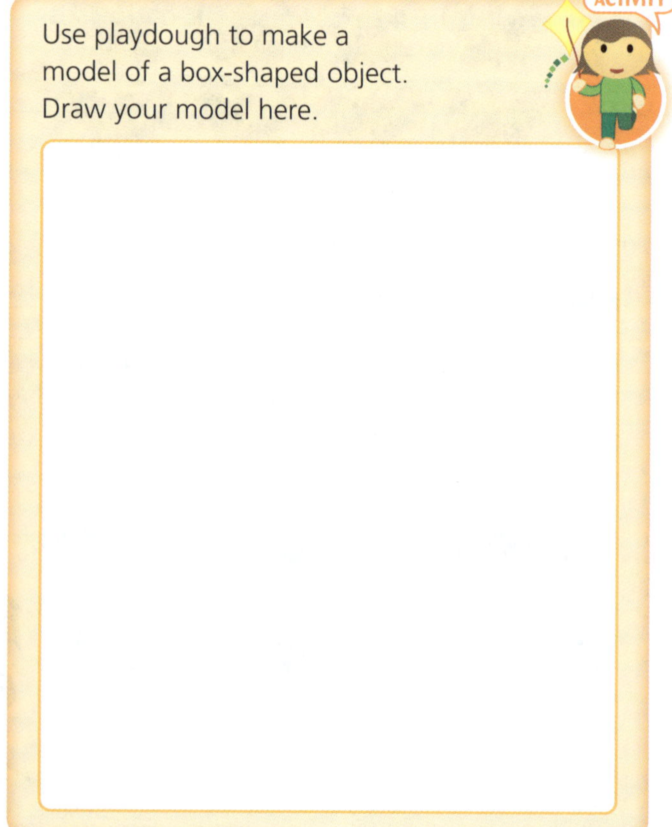

Length

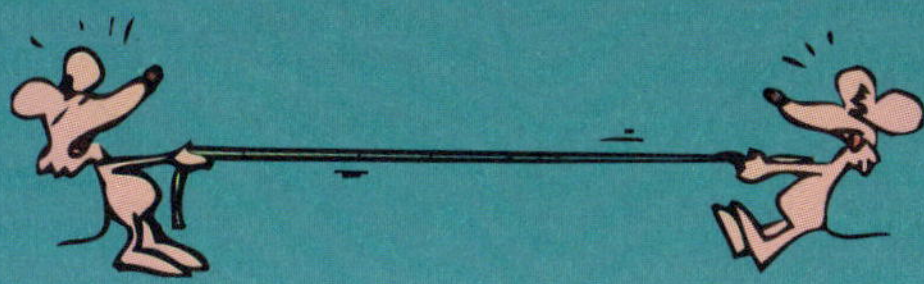

Discuss this picture.
Draw a circle around: the lowest kite, the tallest bush, the highest bird, the longest snake, the longer truck, the house further away, the flower to the right of the bin, the echidna closer to the bin, the duck halfway across the pond, the cats side by side, the shortest dog, the lowest cloud, the flowers between the rocks, the cars parked end to end, the flower to the left of the cow.

23A Rearranging groups

CONCEPT

Is the number of cakes in the first picture the same as the number in the second?

How many cakes? ☐

How many cakes? ☐

Moving objects does not change how many there are.

1 Match equal groups by drawing lines.

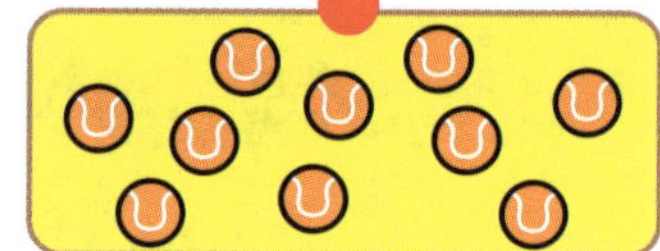

2 Match equal groups by drawing lines.

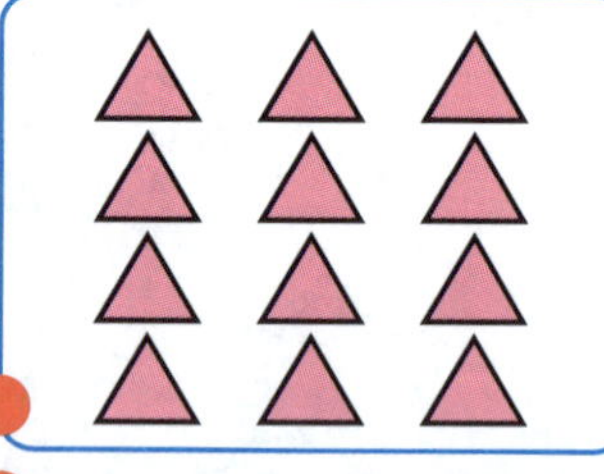

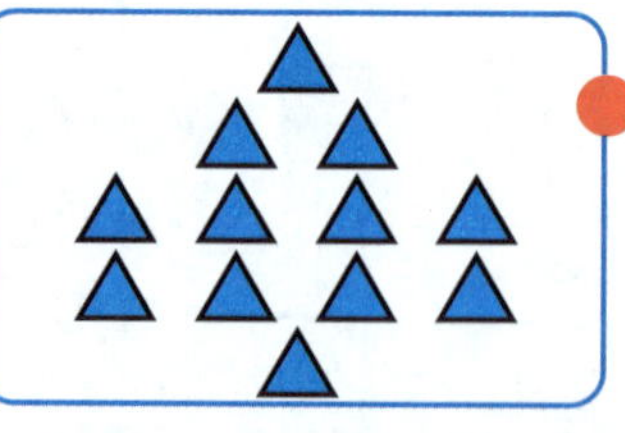

INVESTIGATION

Take a handful of counters. Have a partner match your counters. Take turns to do this.

 • *AUSTRALIAN SIGNPOST MATHS NSW K* • ISBN 9780655709015

23B Using grouping to share

CONCEPT

How many children can be given 3 counters?

15 counters

I have circled groups of 3.

There are 5 groups of 3 in 15 counters.

5 children can be given 3 counters each.

Use counters to model each question.

1 How many groups are there? Circle the groups.

a 12 marbles, 4 in each group. ☐ groups

b 8 bottles, 2 in each group. ☐ groups

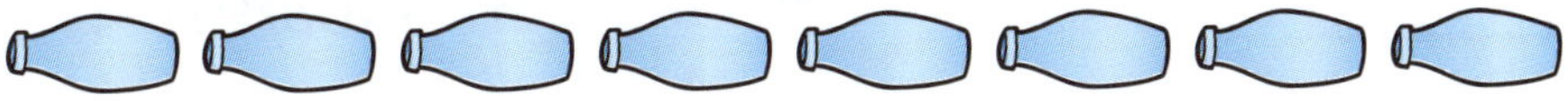

c 15 counters, 5 in each group. ☐ groups

2 Here we have groups of 3 toys.

a How many groups of 3 are there altogether? ☐ groups

b How many toys are in 4 groups of 3 toys? ☐

c How many toys are in 3 groups of 3 toys? ☐

3 Use these groups to find how many counters are in:

a 5 groups of 2 ☐

b 8 groups of 2 ☐

 • *AUSTRALIAN SIGNPOST MATHS NSW K* • ISBN 9780655709015

1 Use colour to match shapes. Discuss the features of each shape.

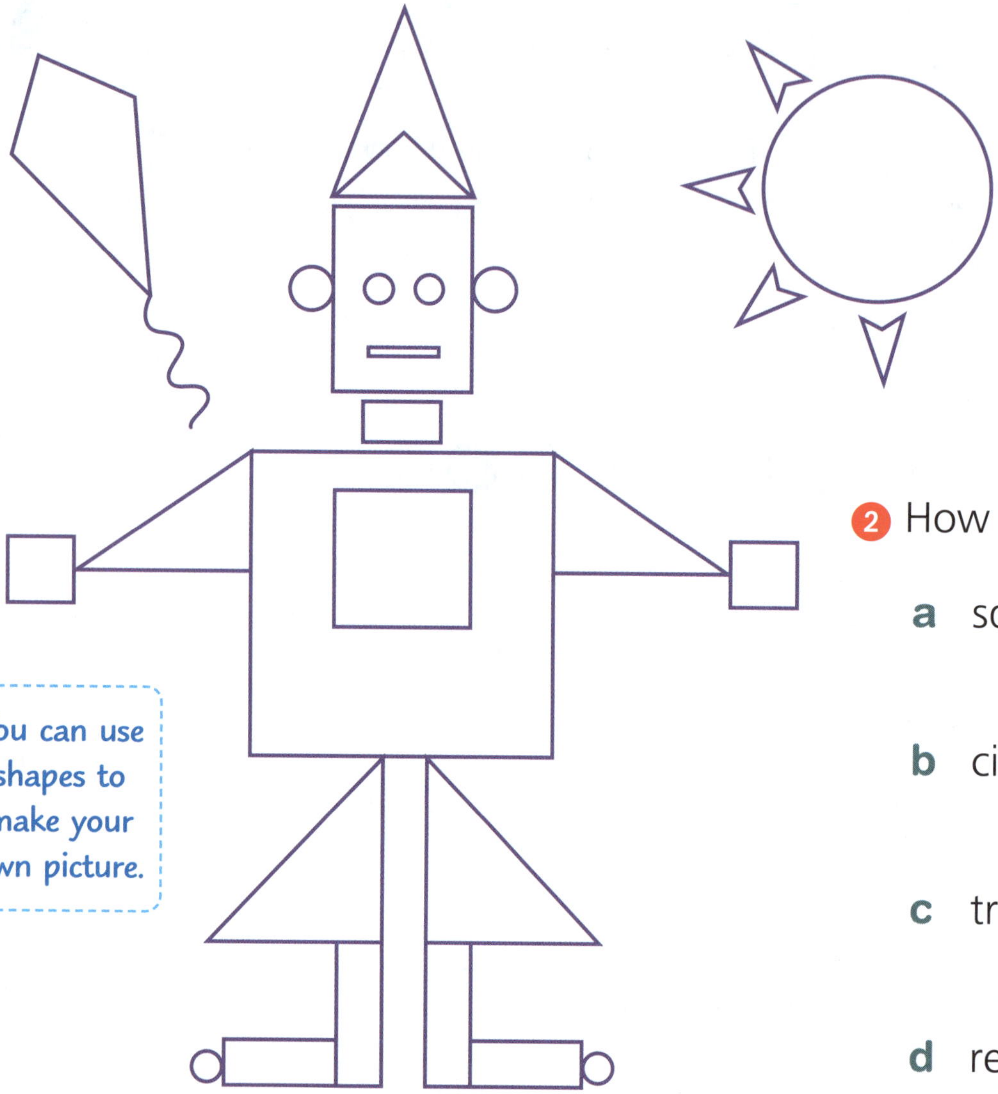

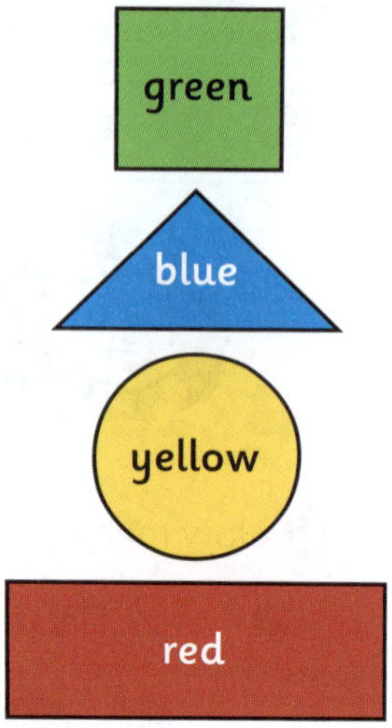

green

blue

yellow

red

You can use shapes to make your own picture.

2 How many:

a squares?

b circles?

c triangles?

d rectangles?

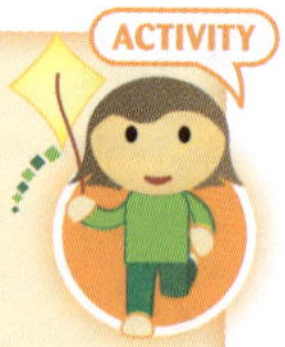

ACTIVITY

Use pipe cleaners to make these shapes. Draw your shapes below.

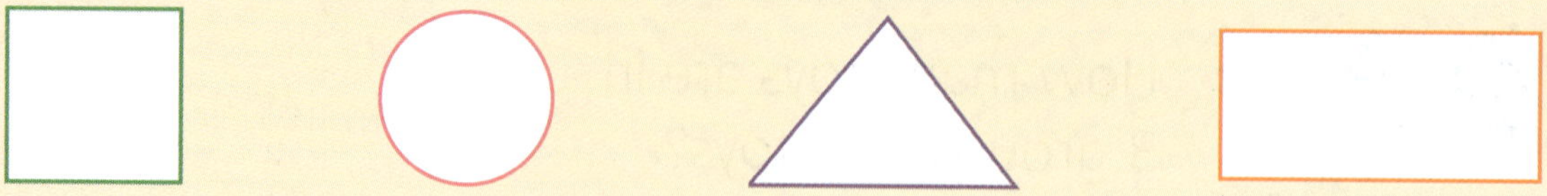

 • *AUSTRALIAN SIGNPOST MATHS NSW K* • ISBN 9780655709015

23D Data displays

1 Use the pictures to complete the graph. Discuss possible responses.

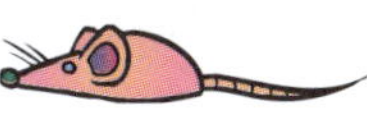

Class Pets

fish	mice	birds

a How many fish? 

b How many mice? 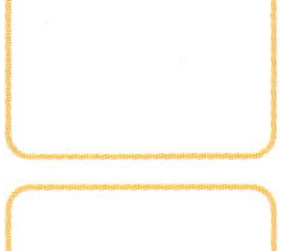

c How many birds? 

d How many pets altogether? 

Collecting information

10 students will be chosen to say if they have a brother. Put a tick next to "yes" or "no" for each student. Discuss the possible results.

Do you have a brother? Yes or no?										
yes										
no										

24A Counting to 30

1

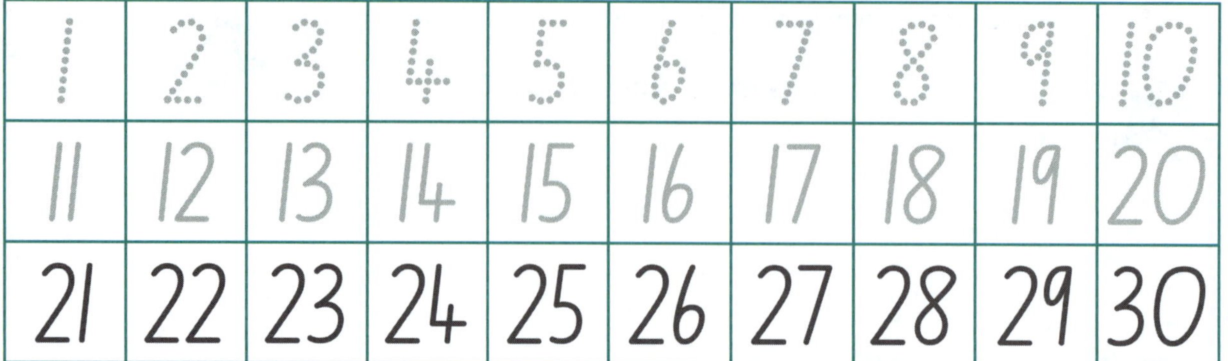

1	2	3	4	5	6	7	8	9	10
11	12	13	14	15	16	17	18	19	20
21	22	23	24	25	26	27	28	29	30

- Point to the numbers as you count them. Count from 1 to 30.
- Start at 30 and count backwards.
- Start at 13 and count to 27.
- Trace over the numbers 1 to 10.
- **Circle** every second number. Say them aloud.

- Colour **red** the numbers that have a **5** in them.
- Colour **blue** the numbers that have a **0** in them.

2 Write the number one more than:

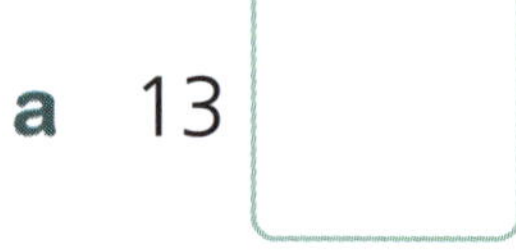

a 13 ☐ **b** 9 ☐ **c** 15 ☐

d 18 ☐ **e** 11 ☐ **f** 17 ☐

13 is one ten and three ones.

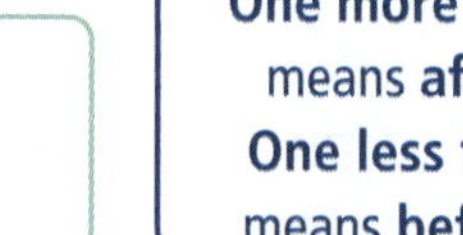

3 Write the number one less than:

a 18 ☐ **b** 13 ☐ **c** 16 ☐

d 15 ☐ **e** 19 ☐ **f** 10 ☐

One more than means **after**.
One less than means **before**.

INVESTIGATION

The Wurundjeri people have a counting system that uses parts of the body to represent numbers.

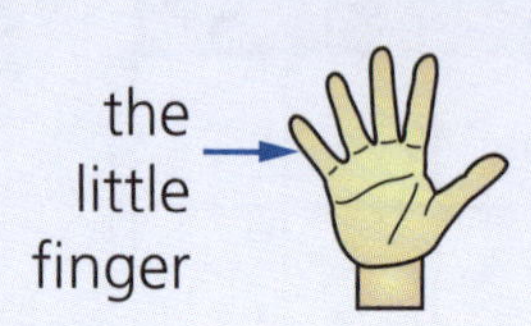

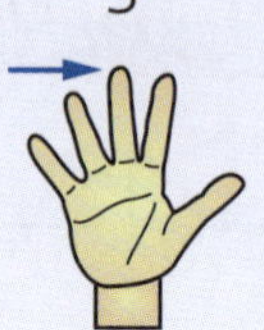

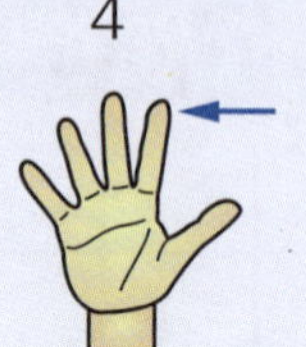

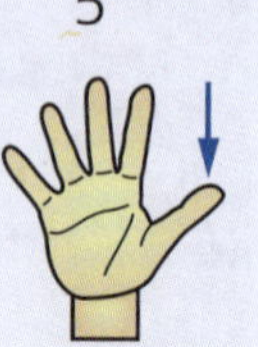

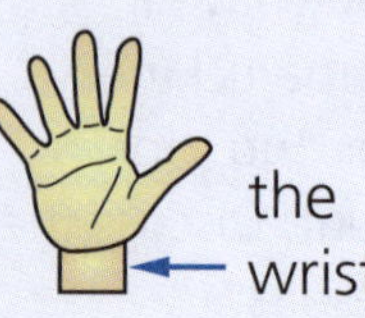

6
the wrist

 • *AUSTRALIAN SIGNPOST MATHS NSW K* • ISBN 9780655709015

Adding on and counting back

1 Use counters to add by counting on. Put extra counters on the line, one at a time.

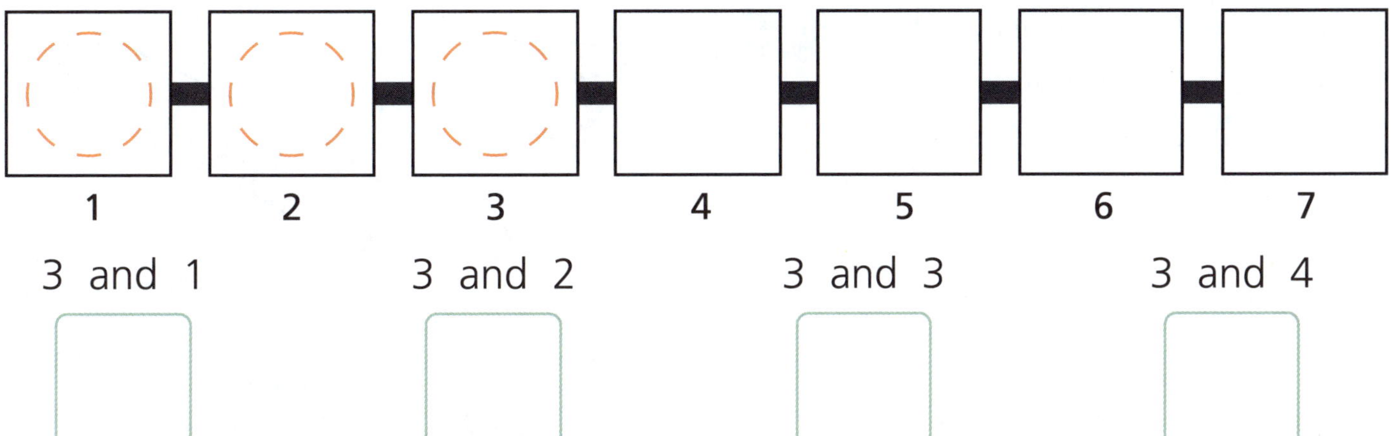

Use the counting line above to subtract, by taking away one counter at a time.

2 Add by counting on, lifting one finger at a time.

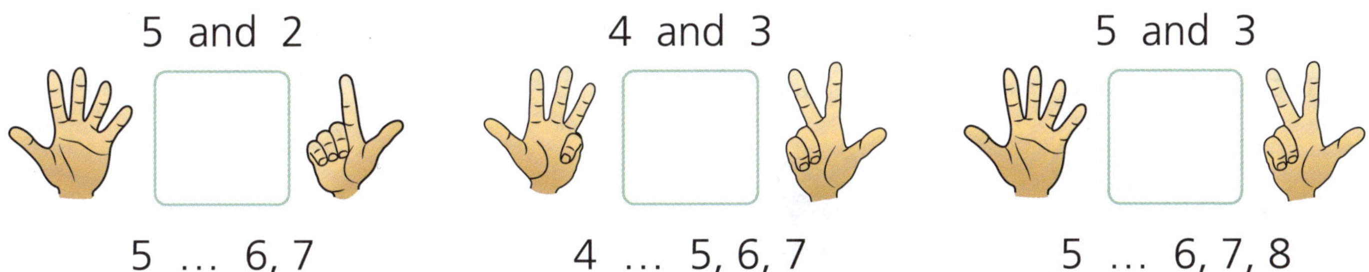

3 Subtract by counting back, putting down one finger at a time.

5 take away 1

5 take away 2

5 take away 3

24C Can-shaped objects

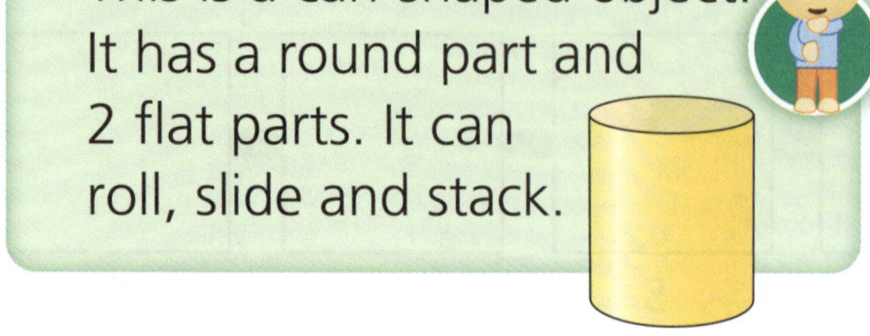

1 Trace this can-shaped object.

3 Circle the can-shaped objects in this picture.

Talk about these objects using the words round, pointy, curved, straight, flat, rolls, slides and stacks.

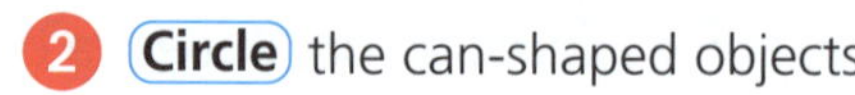

2 Circle the can-shaped objects.

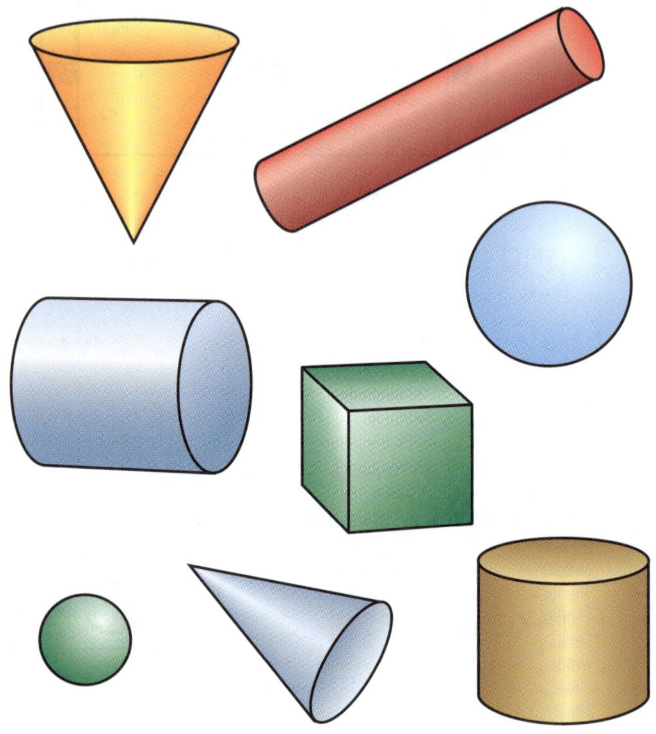

FUN SPOT

Use playdough to make a model of a can-shaped object. Draw your model here.

 • *AUSTRALIAN SIGNPOST MATHS NSW K* • ISBN 9780655709015

Lighter or heavier?

1 **Circle** the objects that are lighter than your lunch box. Explain your reasons.
What other objects would be about the same weight as the apple?

ACTIVITY

Draw pictures to match.

I weigh about the same as a...

lighter than my book	heavier than my book

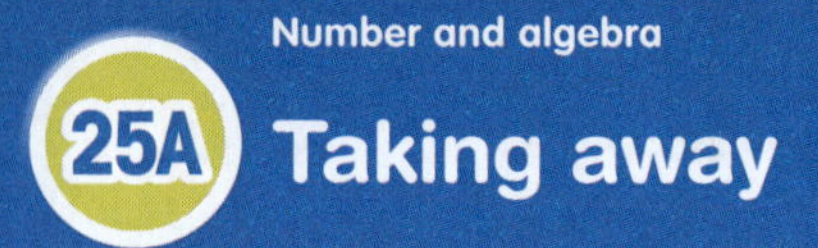

Taking away

4 take away 2 leaves 2.

1 How many are left?

a 

9 take away 3 leaves  cows.

b 

8 take away 3 leaves  coins.

c 

10 take away 5 leaves  fish.

2 Draw counters to find the answers. Talk about these.

a 6 take away 4 leaves 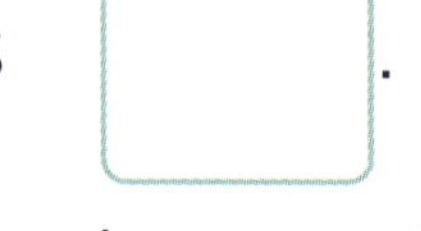.

If you put back 4 what happens?

b 10 take away 7 leaves .

If you put back 7 what happens?

Taking away

1 Cross off three objects in each row. How many are left?

a

b

c

Discuss what happens when the objects are added back on.

2 Cross off four objects in each row. How many are left in each row?

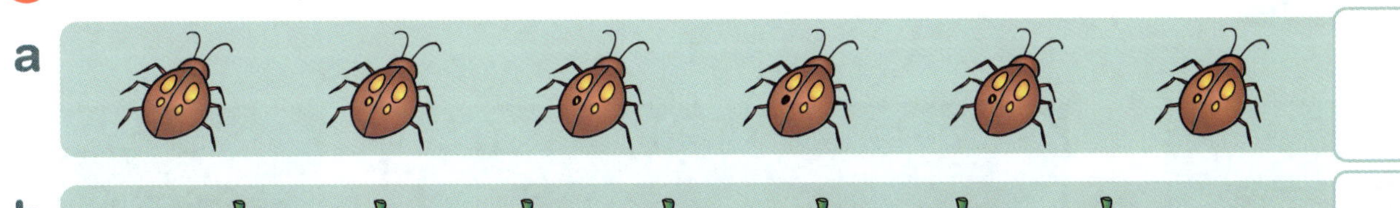

a

b

c

d

Draw and complete.

- 9 fish, take away 5.

How many are left?

- 8 balls, take away 6.

How many are left?

25C Giving and following directions

1 Trace each path with your finger.
To go to **D** your finger could pass through 2, 6 and 7 or 1, 3 and 7

Begin at **Start**. Write the numbers your finger passes through to reach:

A ______ **B** ______ **C** ______ **D** ______

E ______ **F** ______ **G** ______

Start

How would you go to the ice-cream shop **H**? ______

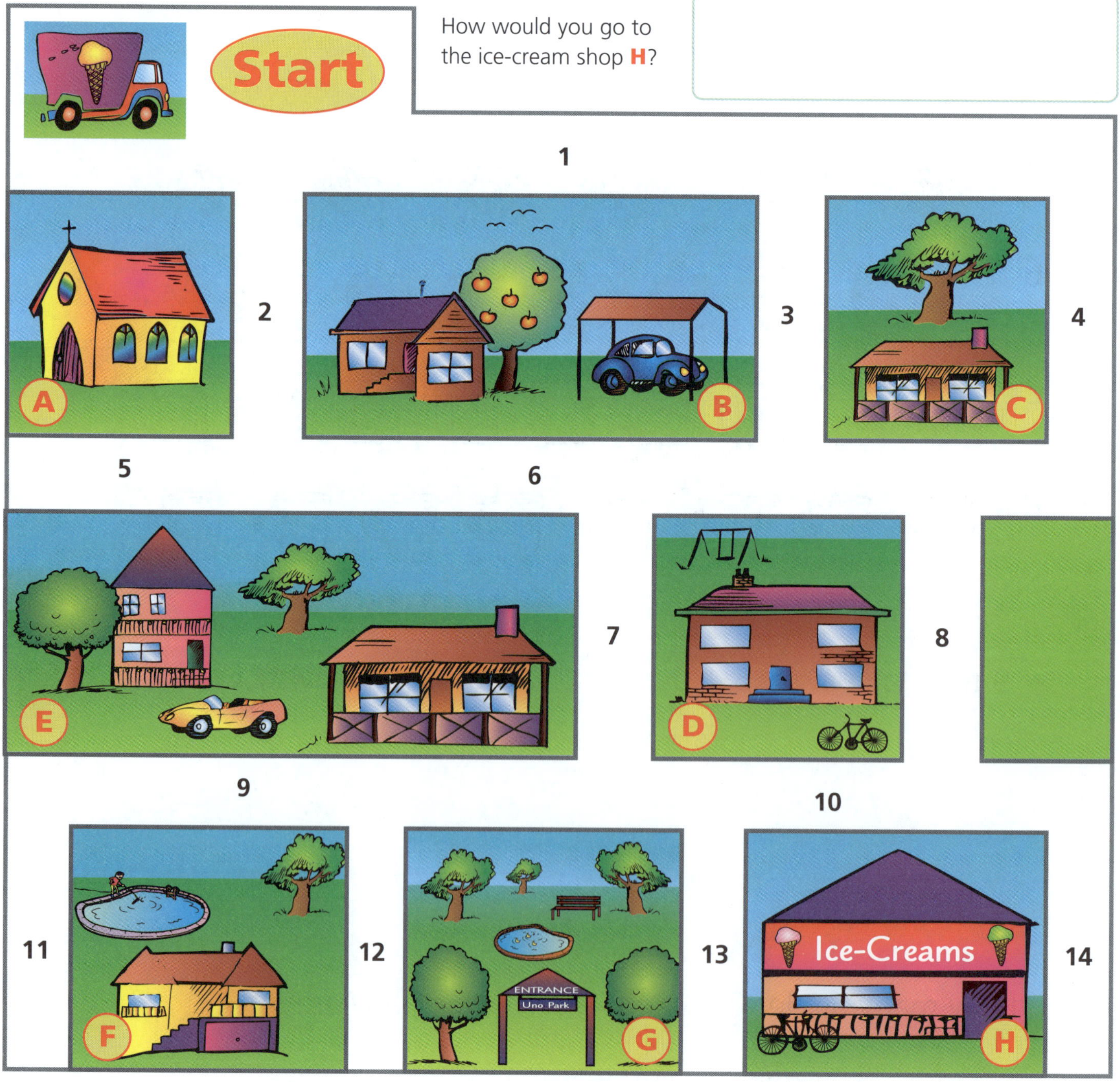

1

A 2 B 3 C 4

5 6

E 7 D 8

9 10

11 F 12 G 13 Ice-Creams H 14

ENTRANCE
Uno Park

25D Sequencing events

1 Draw lines to show the order.

first

2 Draw something you do before school.

Draw something you do after school.

before

after

3 Read the days of the week. Fill in the empty boxes. Colour the school days. How many days in a week?

Sunday	Monday	Tuesday	Wednesday	Thursday	Friday	Saturday
1st	2nd					

What day is: the 2nd day of the week? the 6th day of the week?

26A How many more?

One more makes them the same.

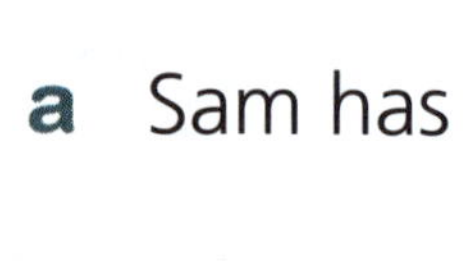

1 Draw more pictures to make the two groups the same.

a Sam has 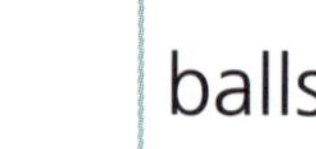 balls.      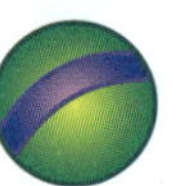 

Jess has  balls.   

We gave Jess ☐ more balls.

How many more did Sam have? ☐

b Lisa has  cakes.     

Sara has  cakes.    

We gave Sara  more cake.

How many more did Lisa have? 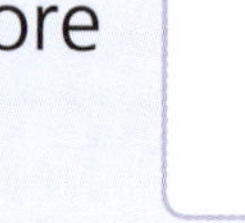

c James has  caps.   

Zen has  caps.     

We gave James  more caps.

How many more did Zen have? 

ACTIVITY

With a partner, compare small groups of objects by lining them up. Find and discuss the difference between each group.

Use groups of objects such as blocks, pencils, beads, counters, craft sticks or toys.

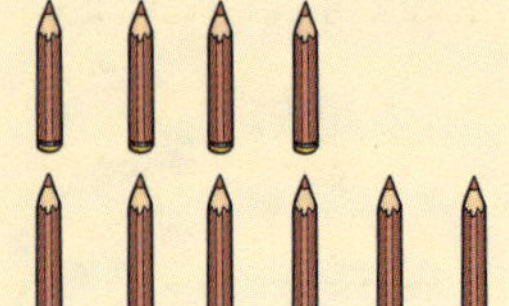

 • *AUSTRALIAN SIGNPOST MATHS NSW K* • ISBN 9780655709015

26B How many more?

Abila took counters from a pile.
Max took the rest.

Abila ●●●●●●●●●●

Max ●●●●●●●●

Max has 8, Abila has 10.
Abila has 2 more than Max.

1 Write the number in each group.
Write how many more are in the larger group.

a

b

c

d
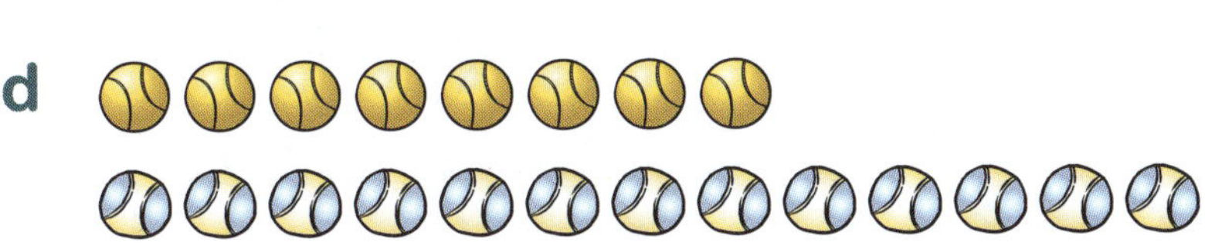

e

f

Place 8 counters in a row.

Place 6 counters in a row.

How many more is 8 than 6? ☐

Use counters to find how many more than 7 is 10. ☐

26C Stacking and packing

 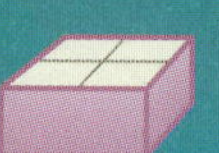 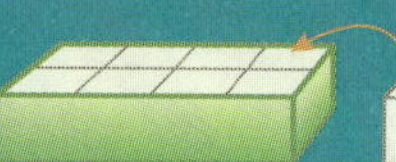 

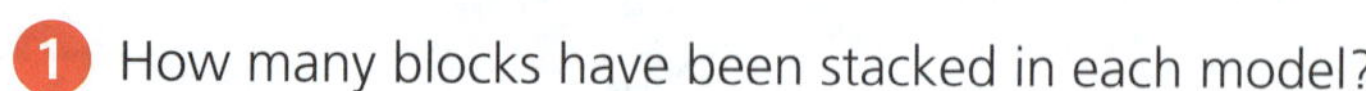

1. How many blocks have been stacked in each model?

Build each model using blocks. Did you use the same number of blocks?

Tick the two models that use the same number of blocks.

2. How many blocks have been packed into each box?

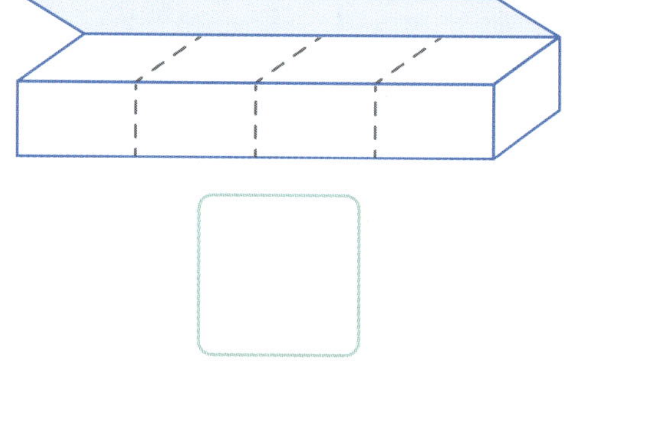

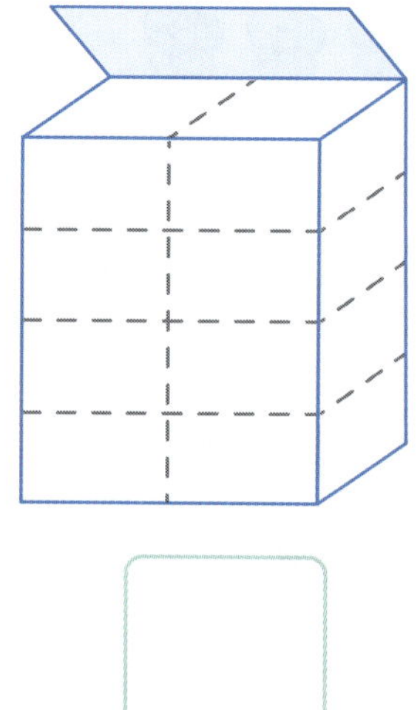

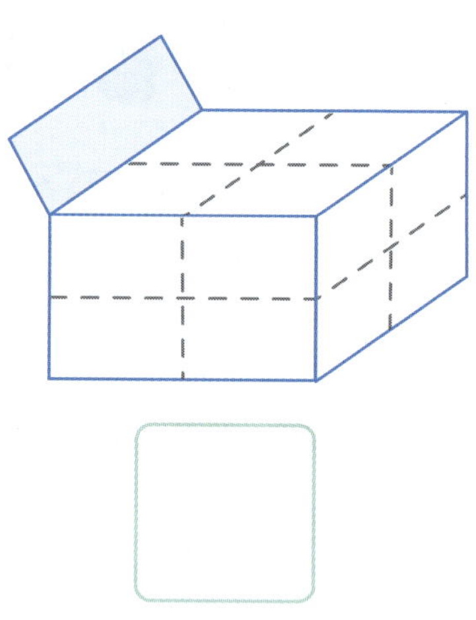

Tick the two boxes that hold the same amount.

Pack blocks into a box.

How many blocks did you use?

Volume

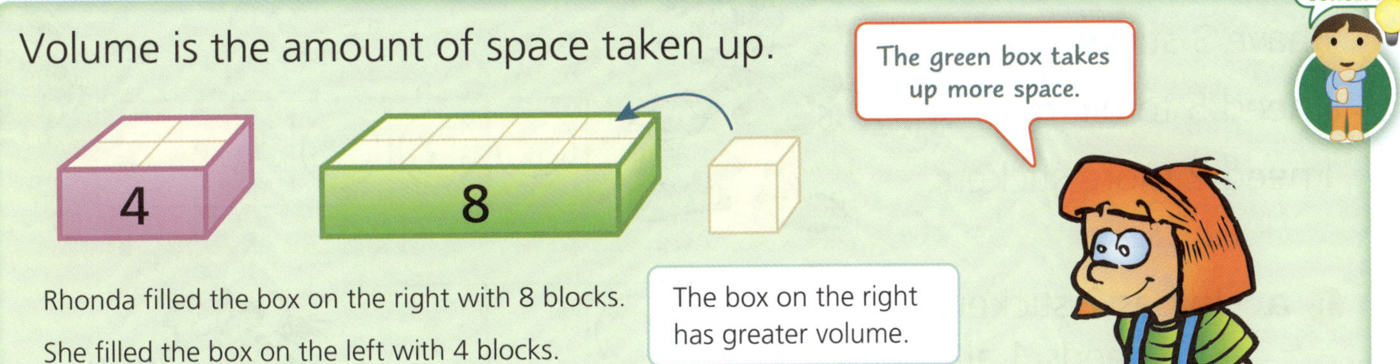

1 Look at boxes A, B and C.

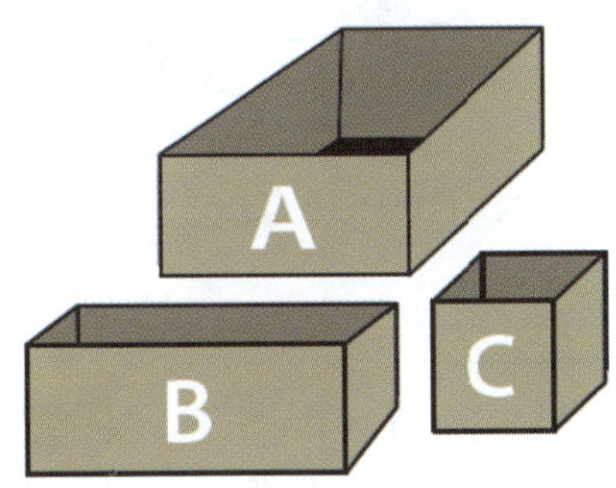

Which box takes up the most space? 

Which box takes up the least space? 

2 **Circle** the blocks that stack easily.

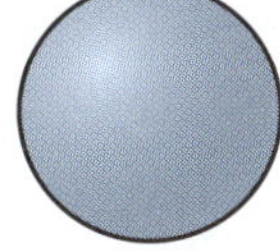
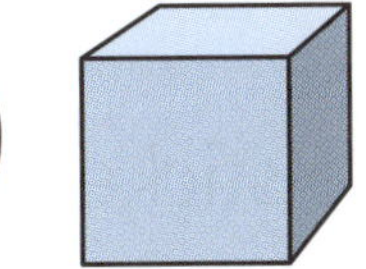
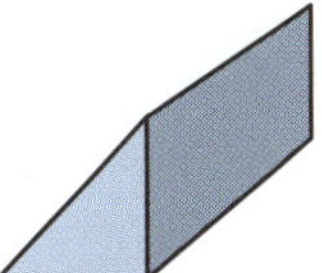
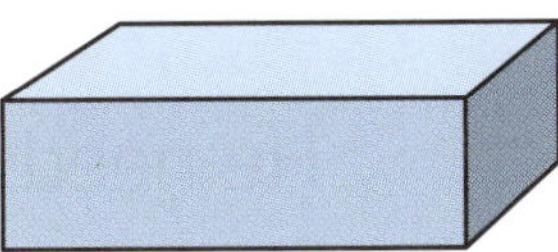

3

A

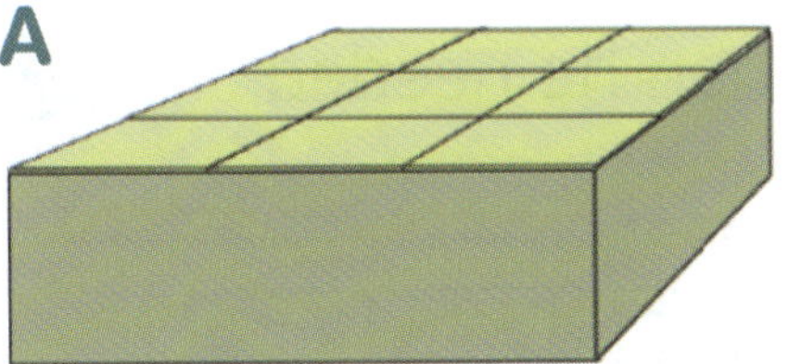

B

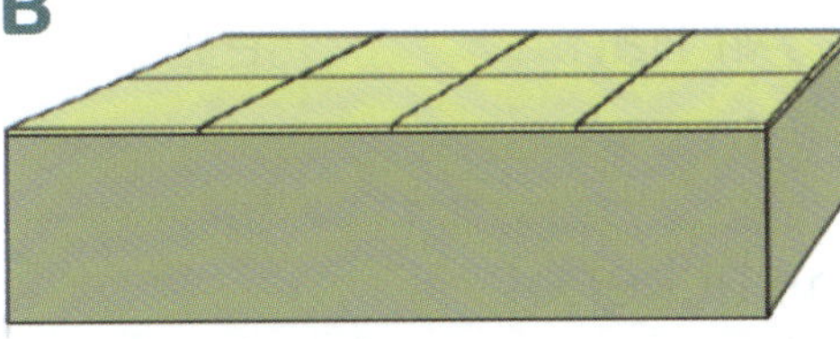

Which has the greater volume?

A holds  blocks.

B holds  blocks.

ACTIVITY

Stack blocks into a container and count how many blocks it takes to fill it. Talk about what you did.

Use blocks to compare the volume of two containers. First, guess which holds the most, then measure.

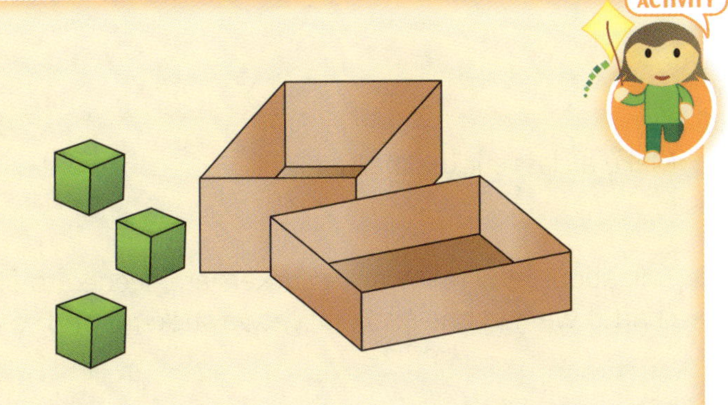

 • *AUSTRALIAN SIGNPOST MATHS NSW K* • ISBN 9780655709015

27A How many more?

CONCEPT

I have 3 stickers.

I need 5 to fill my chart.

I need 2 more stickers.

1 **a** Jo has 2 stickers.
She needs 4 altogether.

She needs ☐ more stickers.

Jo had ☐ less stickers than she needed.

b Ali has 4 stickers.
He needs 5 altogether.

He needs ☐ more sticker.

Ali had ☐ less sticker than he needed.

c Sam has 6 stickers.
She needs 8 altogether.

She needs ☐ more stickers.

ACTIVITY

Make a class sticker chart.

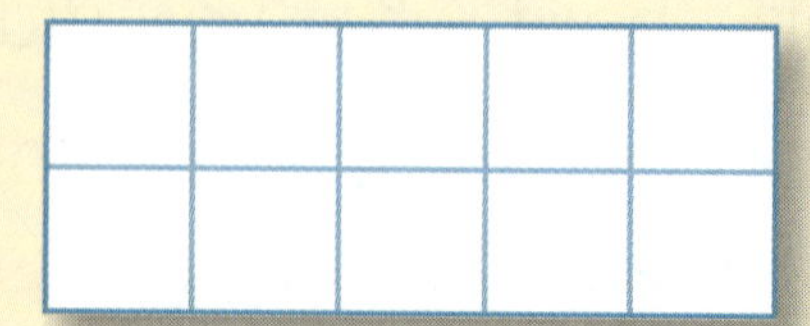

Talk about how many more stickers are needed to fill the chart if we had one sticker? … two stickers? and so on.

Sharing

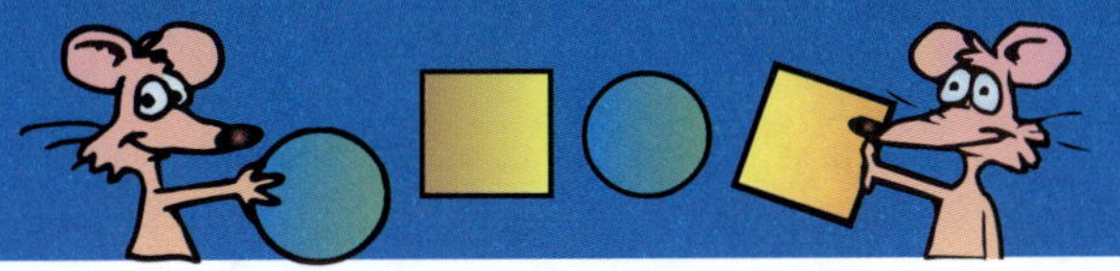

ACTIVITY

1. Use these boxes to share counters between 2 people.

a Start with 8 counters. Share these between 2 people.

How many is each given? ☐

In a fair share, each person gets the same number of objects.

b Start with 12 counters. Share these between 2 people.

How many is each given? ☐

How many have been shared? ☐

2. Use these boxes to share counters among 3 people.

a Start with 15 counters. Share these among 3 people.

How many is each given? ☐

b Start with 12 counters. Share these among 3 people.

How many is each given? ☐

How many have been shared? ☐

27C Classifying 2D shapes

1 Colour the matching shapes in each row.

2 Complete:

Sides can be straight or curved.

a A square has ☐ straight sides.

b A triangle has ☐ straight sides.

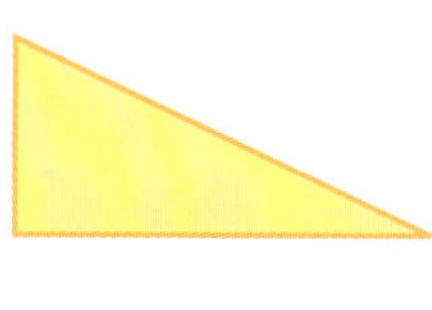

c A rectangle has ☐ straight sides.

d Is the side of a circle straight or curved? ☐

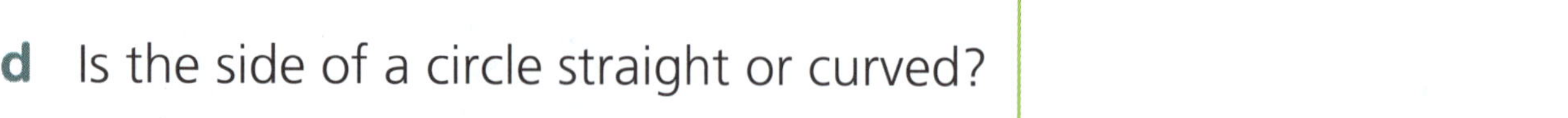

INVESTIGATION

Making shapes

Use craft sticks to make shapes.

How many sticks do you need to make a:

- square? ☐
- triangle? ☐

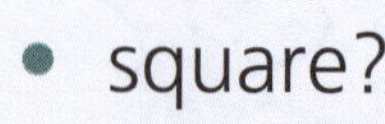

Can you make a circle? yes no Discuss.

A square has 4 sides. A triangle has 3 sides.

 • *AUSTRALIAN SIGNPOST MATHS NSW K* • ISBN 9780655709015

27D Days of the week

There are 7 days in a week. Practise saying the days.

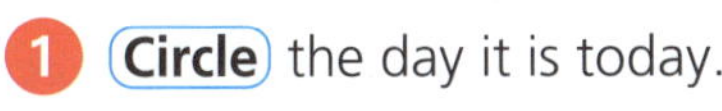

1 Circle the day it is today.  Colour the school days red. Draw lines to join each picture to a day.

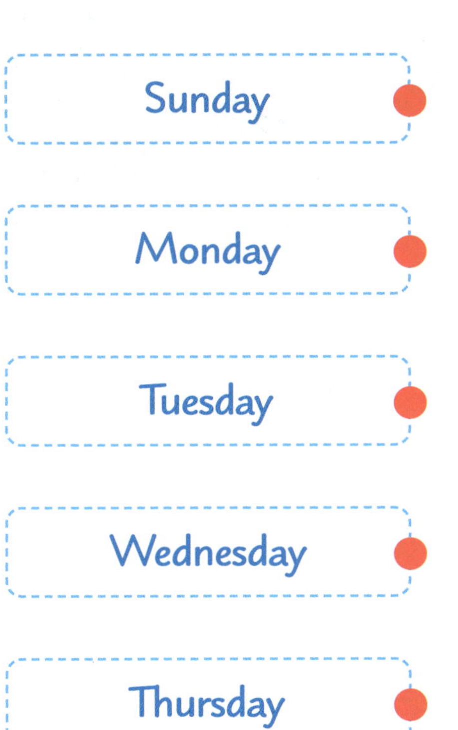

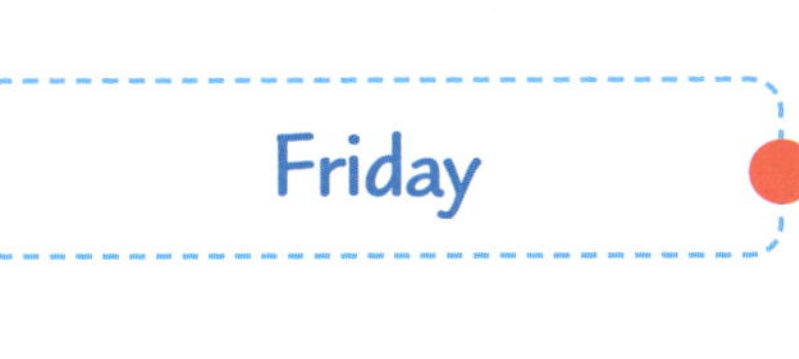

2 Draw something you did yesterday.

3 Draw something you might do tomorrow.

 • *AUSTRALIAN SIGNPOST MATHS NSW K* • ISBN 9780655709015

28A Sharing

CONCEPT

Share these yellow balls between the two boxes.
Share them one at a time until all counters are used.

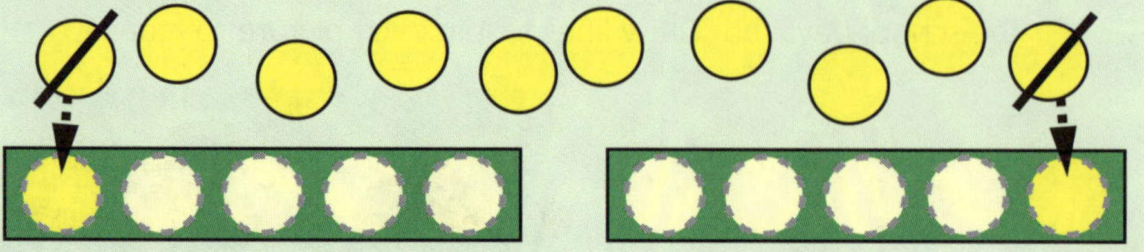

Sharing 10 balls, each box will have 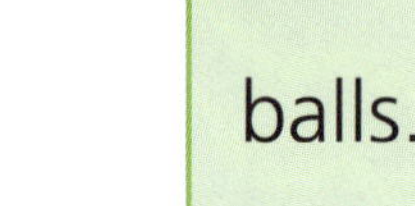 balls.

1. Share these balls into 2 groups. Draw each ball as it is shared.

How many are in each share?

a 6 balls are shared.

b 4 balls are shared.

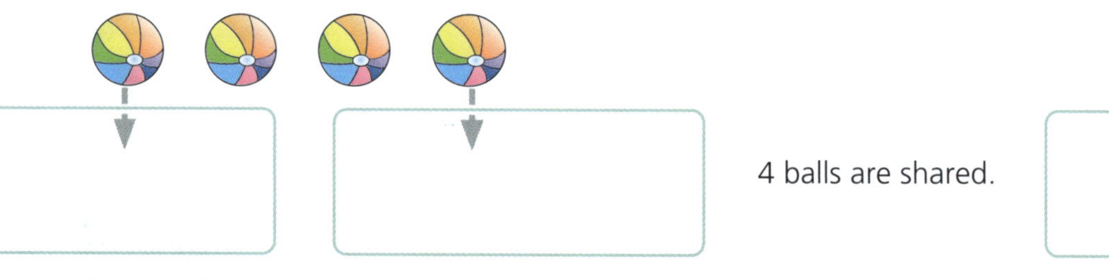

c 8 balls are shared.

d 10 balls are shared.

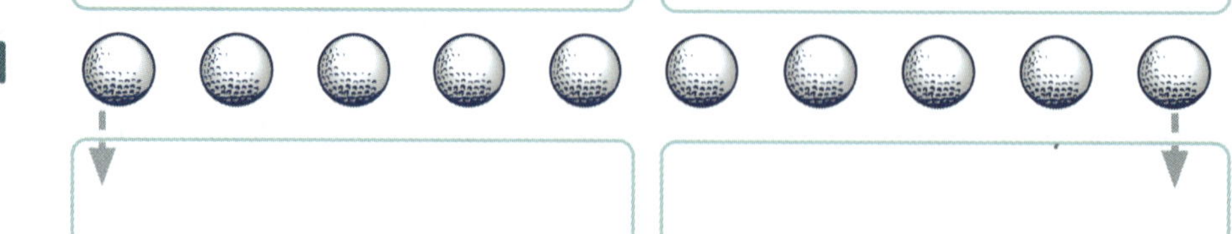

ACTIVITY

Take turns to share counters with a friend.

Take a pile of counters and share them one at a time.

If there is a counter left over, put it aside.

 • *AUSTRALIAN SIGNPOST MATHS NSW K* • ISBN 9780655709015

Sharing among 3 or more

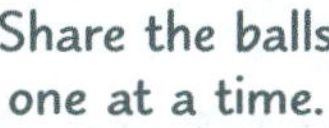

Share the balls one at a time.

Share these red balls among the three boxes.

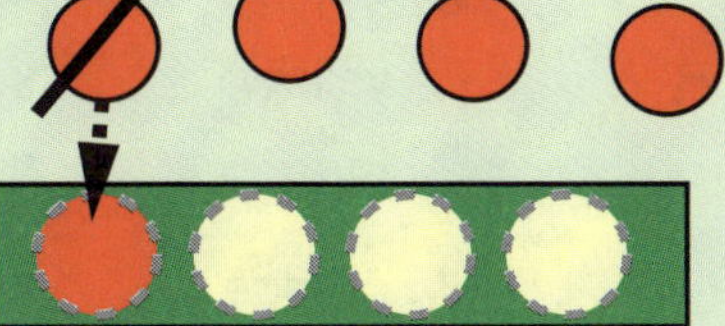
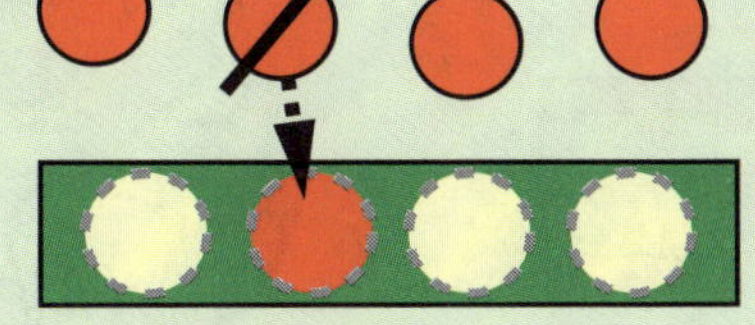
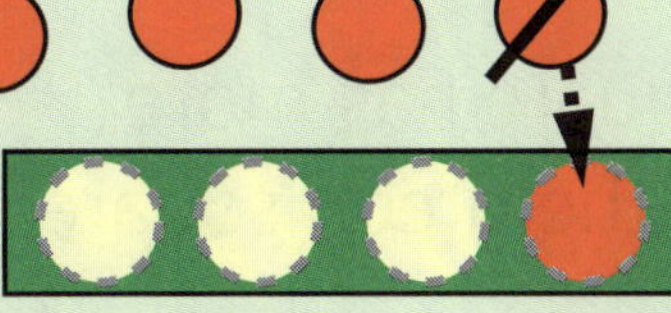

Sharing 12 balls among 3 boxes, each box will have  balls.

1 Share these balls into 3 groups. Draw each ball as it is shared.

a

How many are in each share?

6 balls are shared. ☐

b

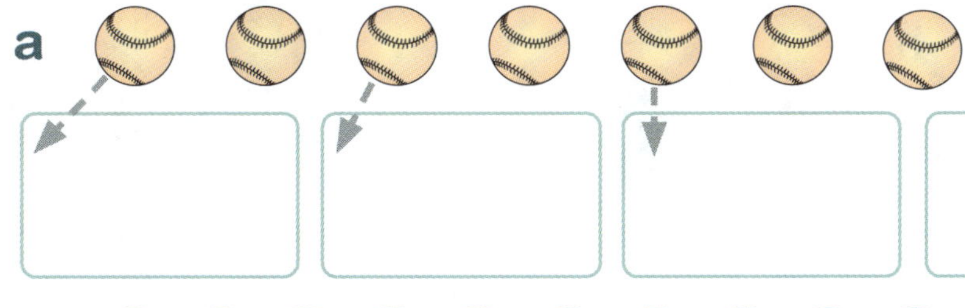

9 balls are shared. ☐

2 Share these balls into 4 groups. Draw each ball as it is shared.

a

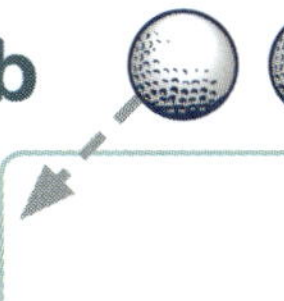
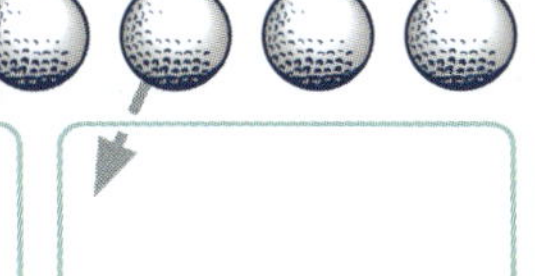

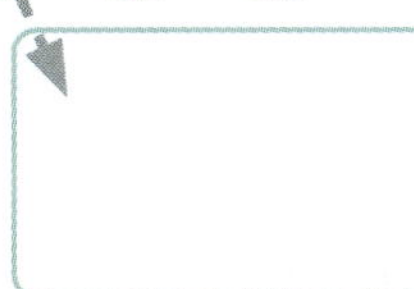

8 balls are shared. ☐

b

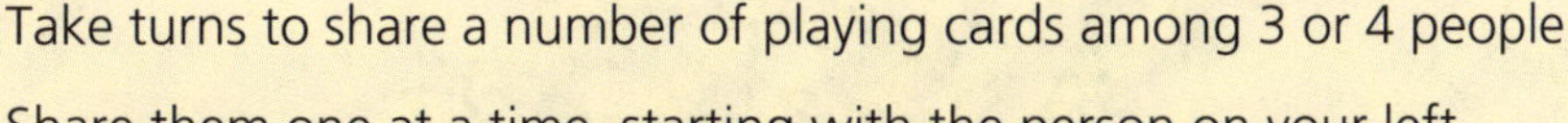

12 balls are shared. ☐

Take turns to share a number of playing cards among 3 or 4 people.

Share them one at a time, starting with the person on your left.

The person with the most picture cards wins.

 • *AUSTRALIAN SIGNPOST MATHS NSW K* • ISBN 9780655709015

28C Comparison of areas

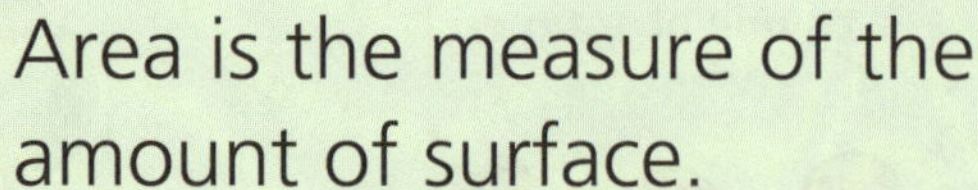

Area is the measure of the amount of surface.

The red square has more area than the green square.

1 Write the letter or colour of the shape with the larger area.

a A B

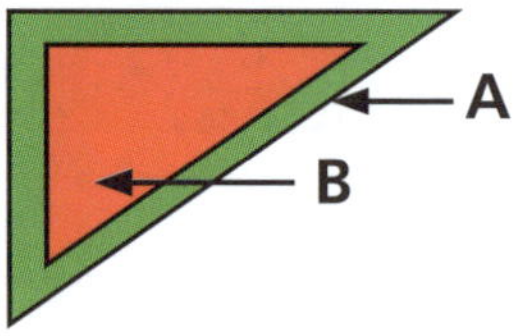

b 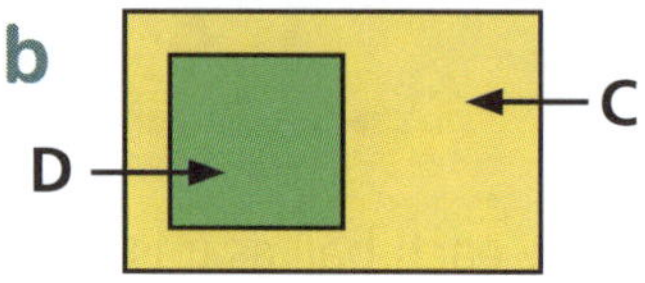 C 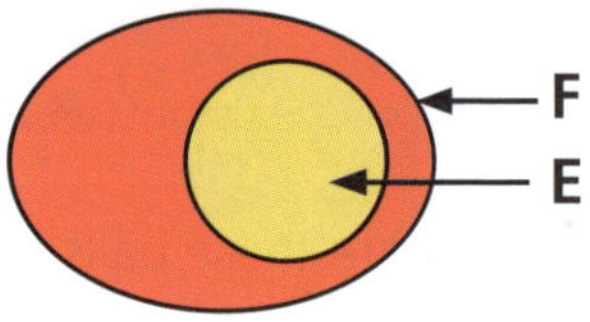 D

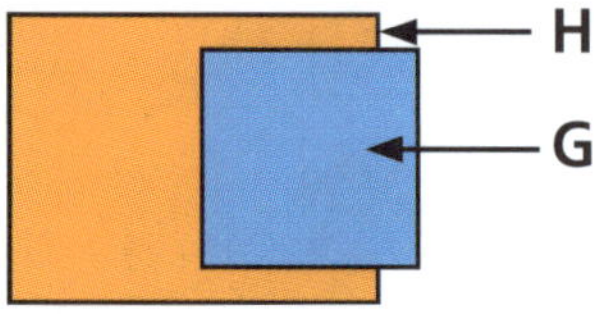

c F E

d H G

e J K

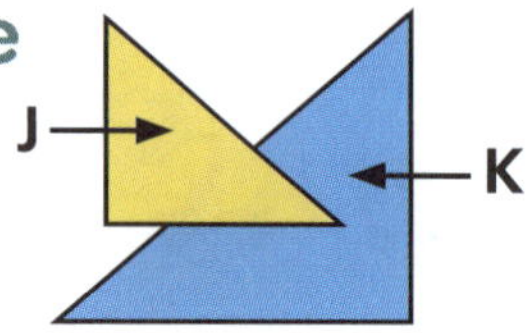

f L M

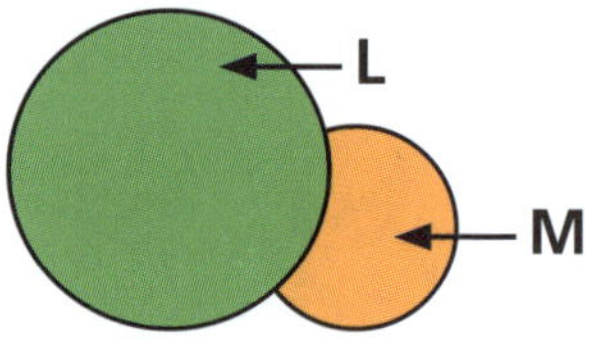

g N P

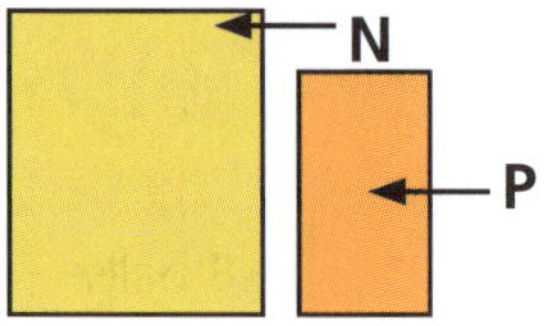

h R Q

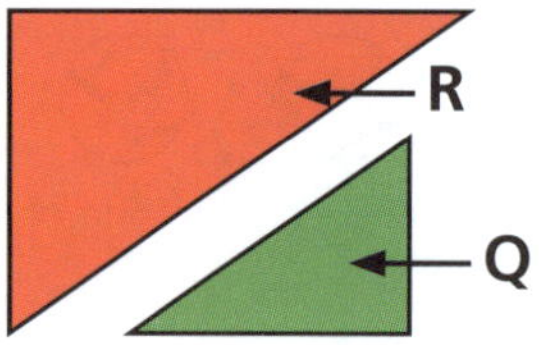

i S T

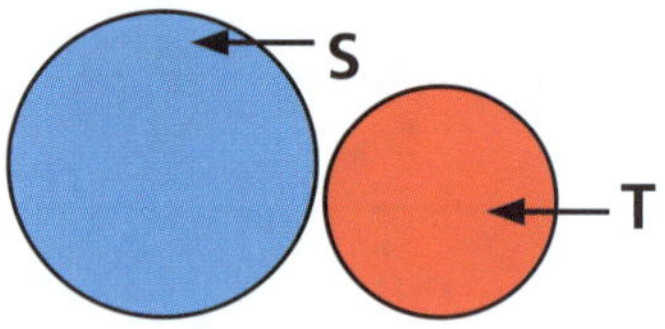

2 Use the letters in question 1 to compare areas.

Choose three flat objects.

By placing one object on top of another, order their top areas from smallest to largest.

 ISBN 9780655709015

28D Comparing areas

Jack placed one piece of paper on top of another to find out which one had the larger area.

The green paper has the larger area.

Then Jack found that the yellow paper was smaller than the red paper. So it must also be smaller than the green paper.

Compare the area of three pieces of paper just like this, putting one piece at the corner of another.

1. Compare the area of these two shapes. Predict which shape has the larger area. Trace shape A onto tracing (or baking) paper. Place the tracing on top of the second shape to find which is larger. Was your prediction correct? Colour the larger shape. Explain why the coloured shape is larger.

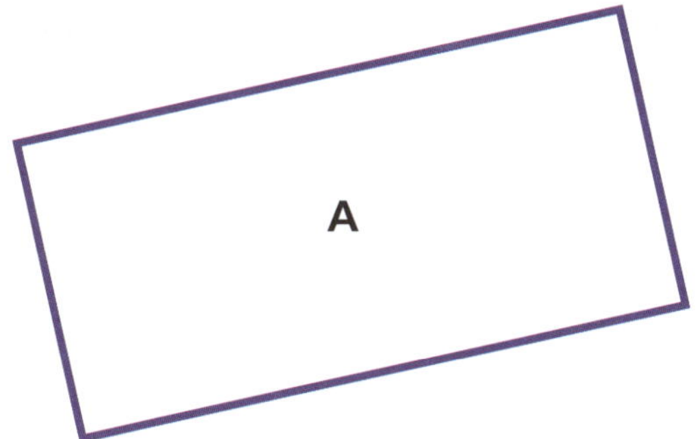

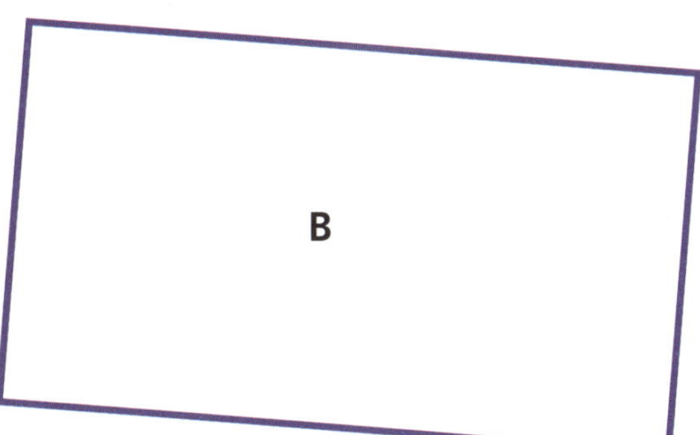

2. Cover shape A with ones blocks.

 blocks covered rectangle A.

Place this book onto other areas in the room to compare the areas.

Draw shapes you found that have a larger area than your book.

Area is the measure of the amount of surface.

Sharing in other ways

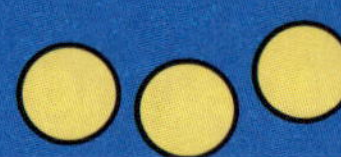

Share these yellow counters between the two boxes.
Share them two at a time until all counters are used.

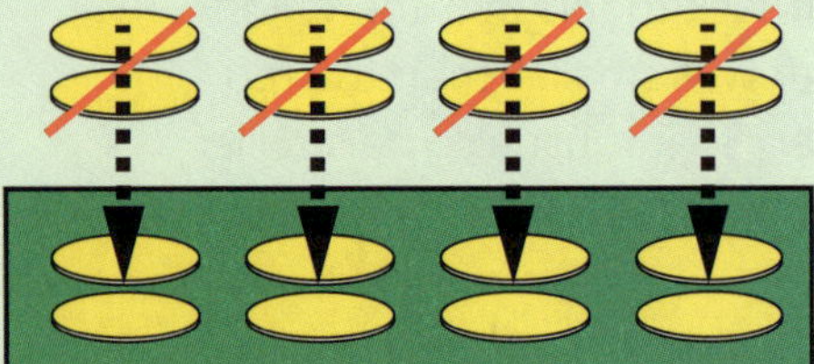
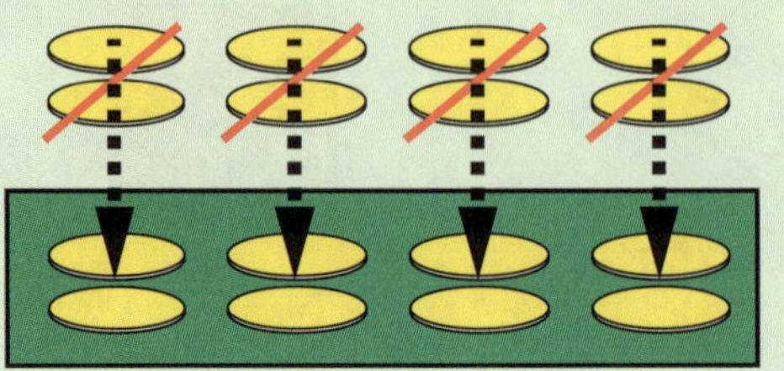

Sharing 16 counters, each box will have  counters.

1. Share two at a time into 2 groups. Draw them as they are shared.

How many are in each share?

a 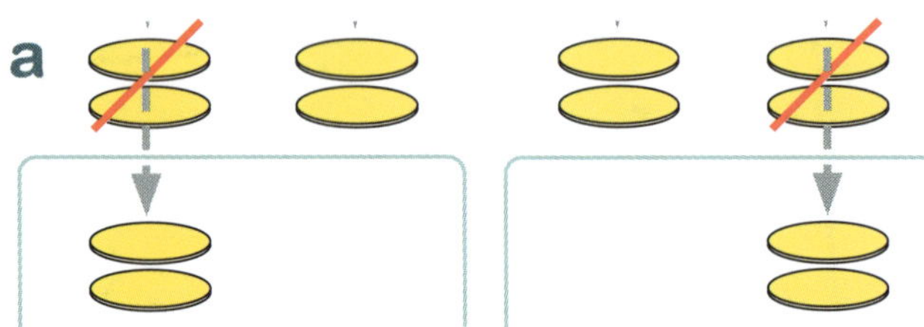

8 counters are shared. 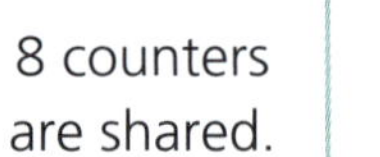

b 

12 cards are shared.

2. Share three at a time into 2 groups. Draw them as they are shared.

a 

18 cards are shared. 

Make two groups of counters, line them up in two rows one row above the other.
Move counters from one group to the other to make two fair shares. Discuss.

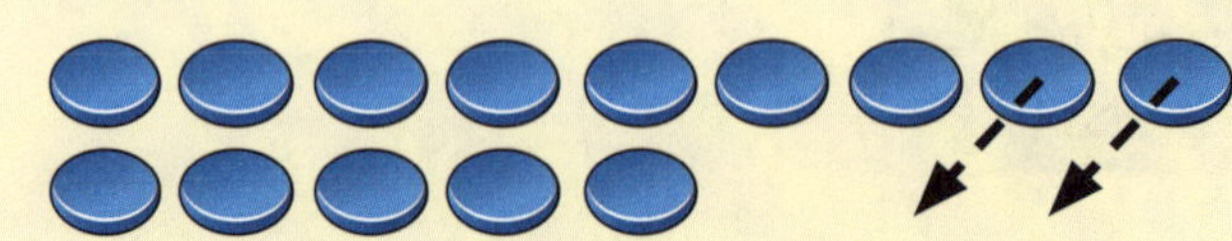

 • *AUSTRALIAN SIGNPOST MATHS NSW K* • ISBN 9780655709015

Comparing lengths

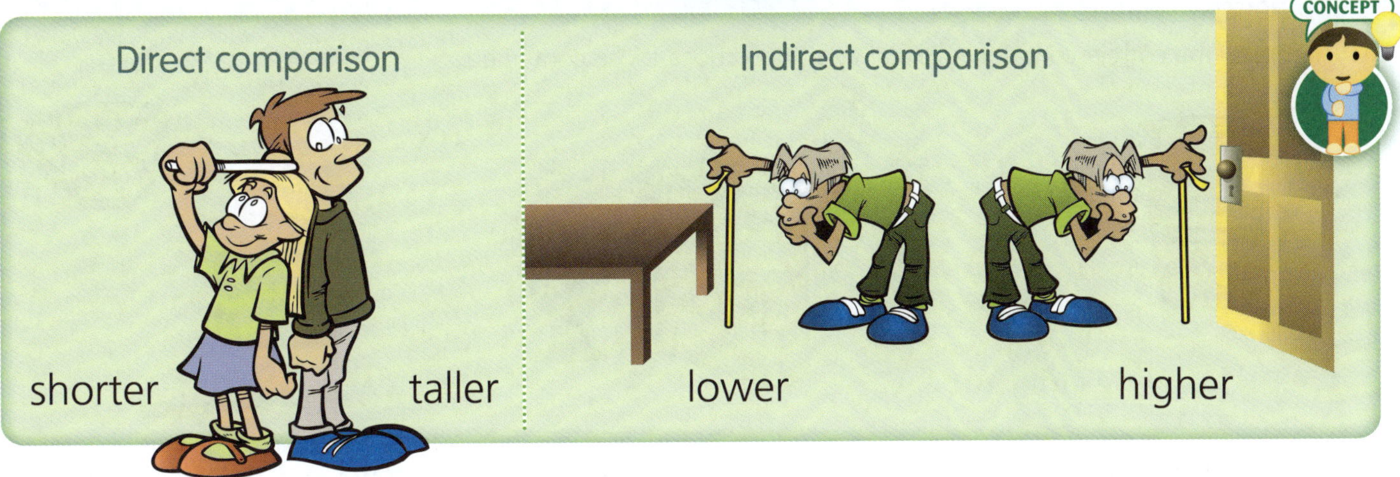

1 Use a piece of string or paper to compare the lengths in each part.
Find out which length is longer and which is shorter.

a A B

☐ is longer. ☐ is shorter.

b C D

☐ is longer. ☐ is shorter.

c E F

☐ is higher. ☐ is lower.

2 ✔ Tick the longest line. ✘ Cross the shortest line.

a

b

Use pieces of string or paper to compare the lengths of three objects in the classroom.

Say which is the longest and which is the shortest and tell someone how you found out.

Discuss whether there are other ways of doing this.

29C Comparing capacities

1 **Circle** the correct containers. Explain why you chose those containers.

holds more | holds less | hold about the same

2 Colour a container that holds more than the glass. Tick a container that holds less than the glass. **Circle** the container that holds about the same as the glass.

3 Each container was filled and then tipped into the same empty jar. Discuss.

Can containers of different shapes hold the same amount?

Which container is about half full?

29D Comparing internal volumes

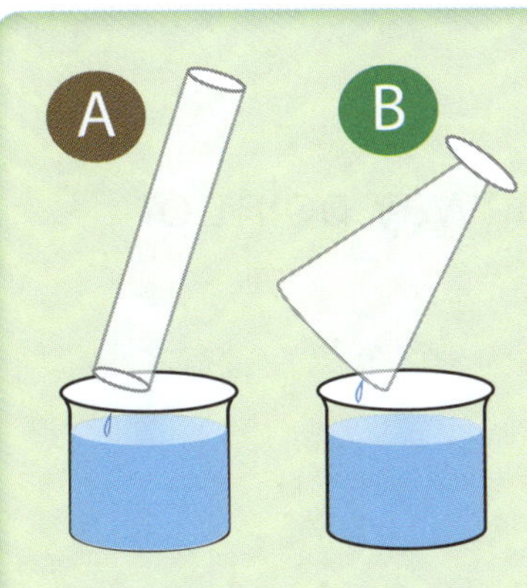

CONCEPT

Volume is the amount of space taken up.
Water and rice have volume because they take up space.

The containers were filled then poured out.

Do A and B hold the same amount?

1 The containers were filled then poured out. **Circle** the container in each question that holds more. **Circle** both containers in the question if they hold about the same amount.

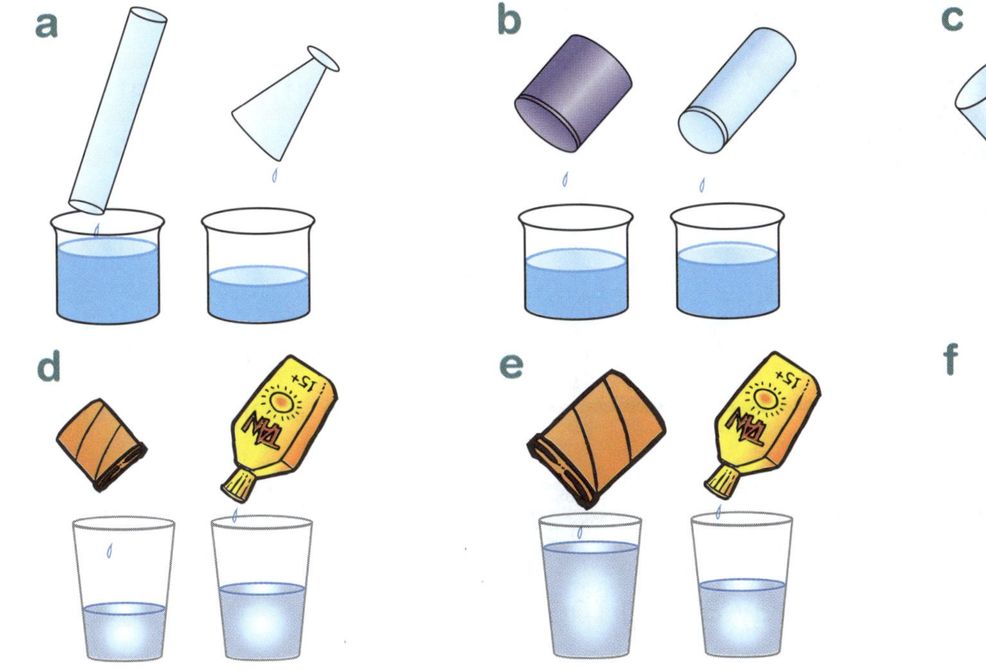

2 Harry has two piles of rice. He wants to find out which pile has the greater volume. He puts each pile into a glass. **Circle** the pile with the greater volume.

Make your own piles. Find out which has the greater volume.
Talk about what you found.

 • *AUSTRALIAN SIGNPOST MATHS NSW K* • ISBN 9780655709015

30A The halfway point

The halfway point is the middle.

1 Circle the halfway point of the see-saw.

2 Circle the halfway point of the rope.

3 Circle the child in the middle of the seat.

4 Is the child with the hat about halfway along the seat?

5 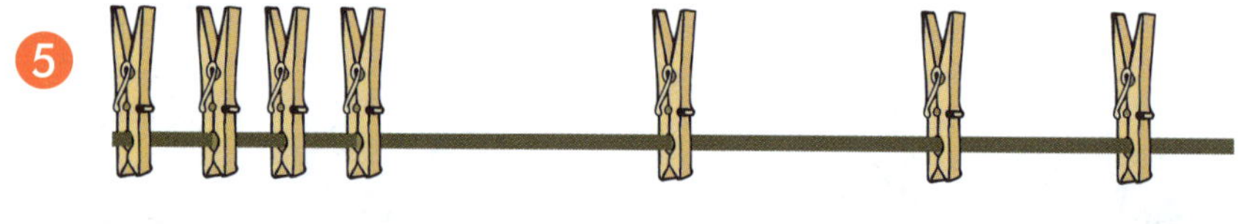

Circle the peg that is halfway along the rope.

6 

Tick the bead in middle.

Circle the bead that is halfway along the string.

Hold the two ends of a piece of string together.

Use a pencil to mark the halfway point.

Cut the string at the halfway mark.

Do the halves have the same length?

Discuss.

30B Halfway

CONCEPT

about halfway

less than halfway

more than halfway

Circle the frog that is halfway through the race. Tick the frogs that are more than halfway. Cross out the one that is less than halfway.

1 Colour the block halfway along the row.

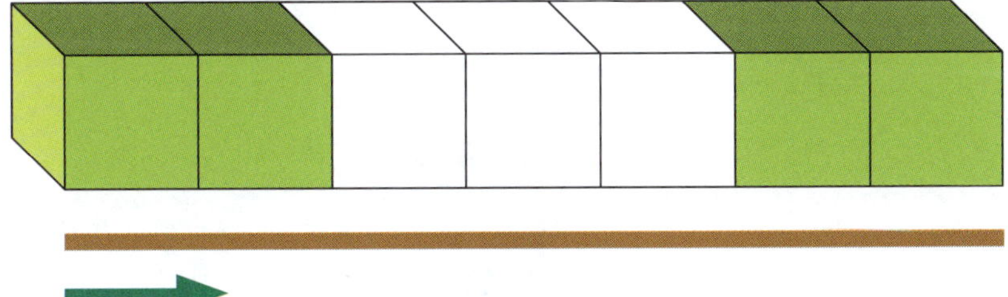

✗ Draw a cross about halfway along this string.

● Draw a dot more than halfway along the string.

✔ Place a tick less than halfway along the string.

2 A — B — C — D — E

What letter is halfway from A to E?

3 0 — 1 — 2 — 3 — 4

Colour the number that is halfway from 0 to 4.

○ Circle the number that is less than halfway from 0 to 4.

✔ Tick the number that is more than halfway from 0 to 4.

 • *AUSTRALIAN SIGNPOST MATHS NSW K* • ISBN 9780655709015

Comparing internal volumes

To find which holds more:

- Fill a container with water or sand.
- Pour it into the other container.
- If it overflows, the first container holds more.

1 In each case, circle the container that holds more. Discuss.

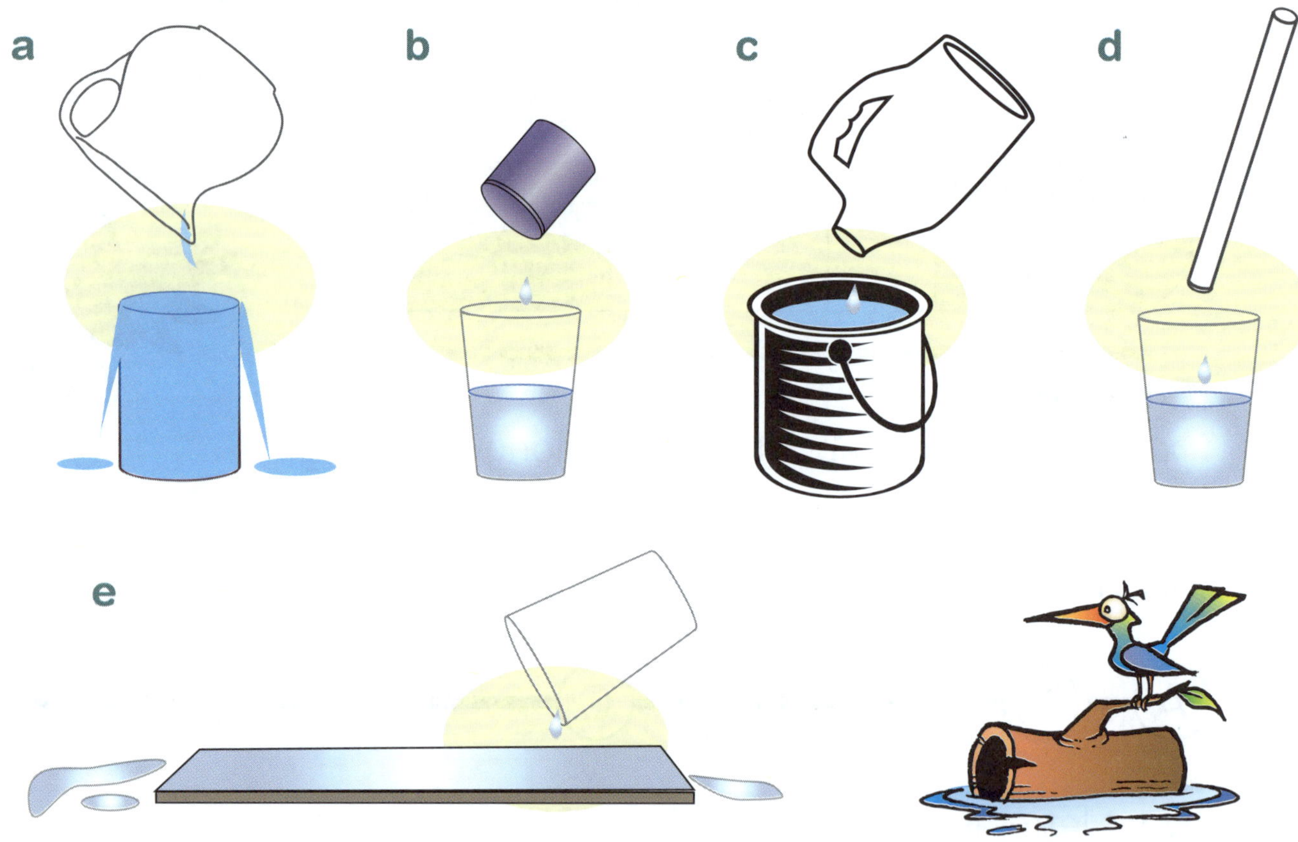

2 Pour water from one container into another to find out which one holds more.
Colour the container that holds more.

30D Volume using blocks

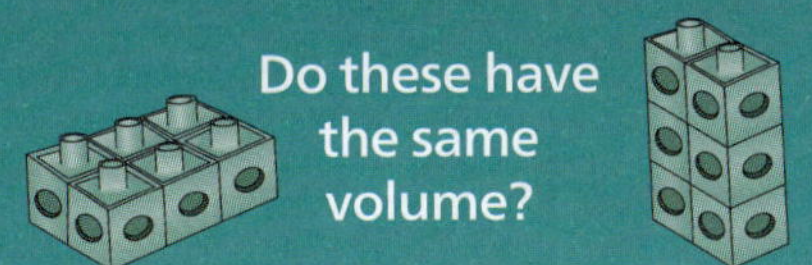

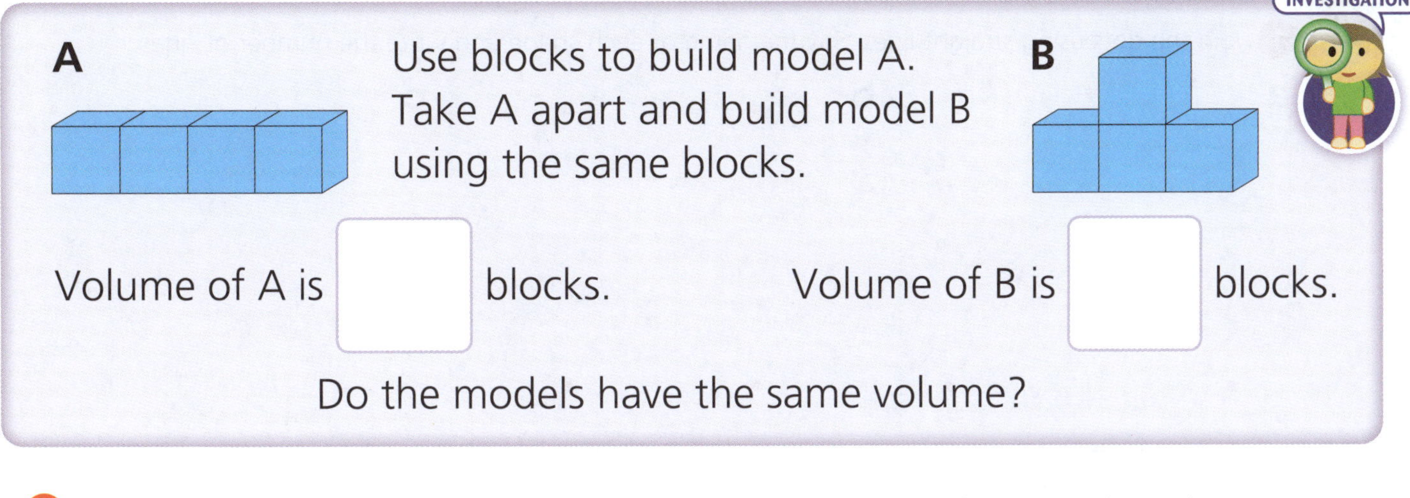

Use blocks to build model A. Take A apart and build model B using the same blocks.

Volume of A is ☐ blocks. Volume of B is ☐ blocks.

Do the models have the same volume?

1 Use blocks or connecting cubes to build each model. Write the number of blocks used in each.

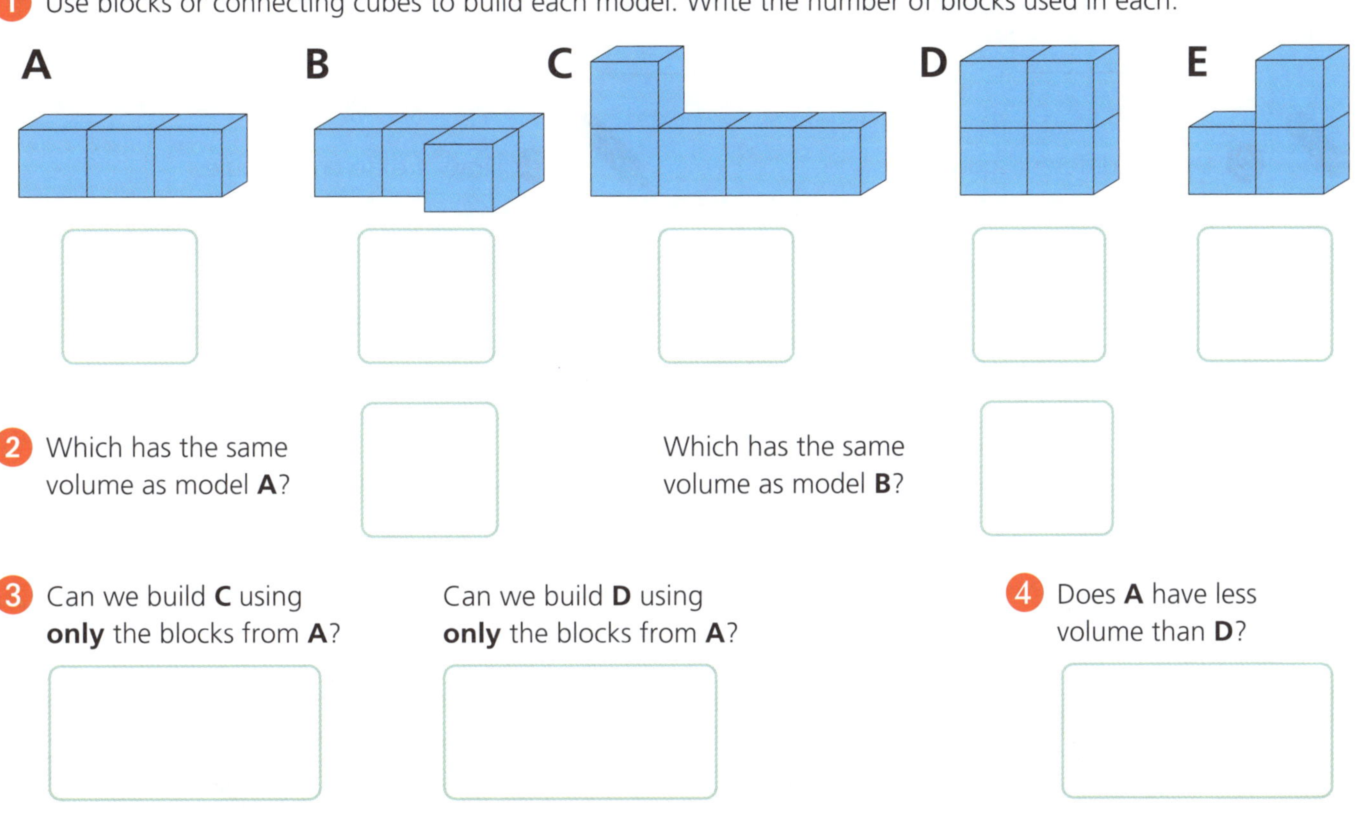

2 Which has the same volume as model **A**? ☐

Which has the same volume as model **B**? ☐

3 Can we build **C** using **only** the blocks from **A**? ☐

Can we build **D** using **only** the blocks from **A**? ☐

4 Does **A** have less volume than **D**? ☐

- Make another model that has the same volume as this model.
- Use 18 blocks to make a model. Your model will take up the same space as 18 blocks.
- Make your own model then count the number of blocks you used.

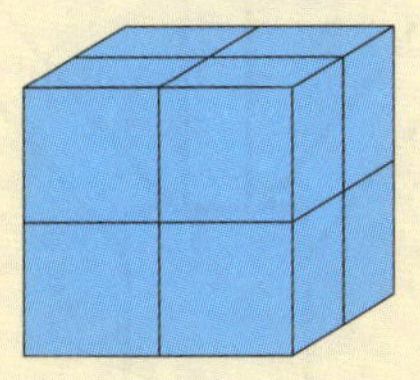

The volume of my model is ☐ blocks.

© PEARSON AUSTRALIA 2023 • *AUSTRALIAN SIGNPOST MATHS NSW K* • ISBN 9780655709015

31A 2D shapes

1 Join the dots using straight lines. Say the name of each shape and write the number of sides.

a ☐ sides

b ☐ sides

c ☐ sides

2 Draw 3 different squares.

3 Draw 4 different triangles.

ACTIVITY

Use playdough, craft sticks, paper or string to make these shape pictures. Make a picture of your own.

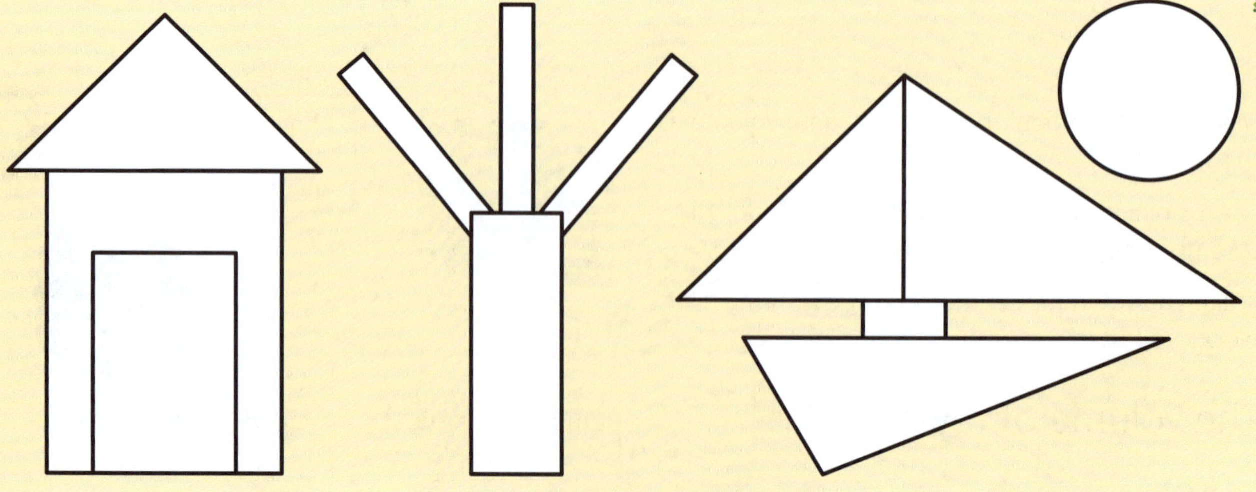

 ISBN 9780655709015

Using data displays

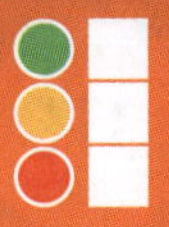

1 Make a display to show the number of boys and girls in your class. Colour a square for each student.

There are ☐ girls.

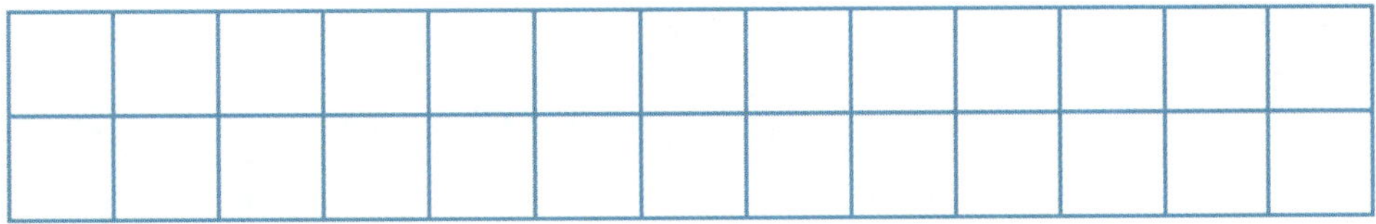

There are ☐ boys.

a How many students are there altogether? ☐

b There are more ☐ than ☐.

ACTIVITY

Talk about this display.
Write a name for this data display.

Ask a friend a question about the display.

31C Describing objects in our world

INVESTIGATION

Discuss the shape of the objects in the picture.

1 Draw objects from the picture above that have:

a a ball shape (sphere)

b 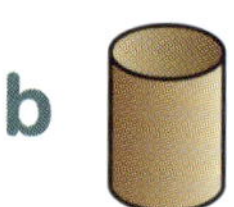 a can shape (cylinder)

c 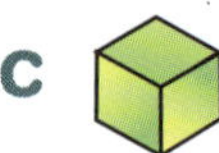 a box shape (prism)

d a cone shape (cone)

Predict then test which of the objects in question 1 will stack.

ACTIVITY

Find ball-shaped, can-shaped, box-shaped and cone-shaped objects in your classroom.

Use playdough to make ball-shaped, can-shaped, box-shaped and cone-shaped objects.

 ISBN 9780655709015

Recording the weather

1 A class made this daily weather chart.

Week 1	Sun (rainy)	Mon (rainy)	Tues (cloudy)	Wed (sunny)	Thurs (sunny)	Fri (sunny)	Sat (sunny)
Week 2	Sun (windy)	Mon (cloudy)	Tues (cloudy)	Wed (windy)	Thurs (sunny)	Fri (sunny)	Sat (rainy)

Discuss the results above and use them to complete this chart.

Weather								Total
sunny								
rainy								
windy								
cloudy								

a How many days were  ? ☐

b How many days were  ? ☐

INVESTIGATION

12 children will be asked what their favourite activity is out of these choices.
Each student first guesses which activity will be chosen most often. Draw a face for each child.

Which of these do children like the most?

swimming										
bike riding										
playing soccer										

32A Location

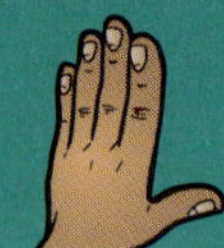

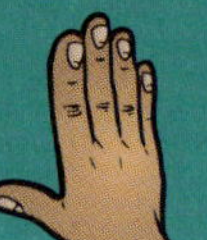

ACTIVITY

Draw:

- a cow inside the paddock
- a duck beside the pond
- birds flying below the clouds
- a house on top of the hill
- a flower under the tree
- yourself in the picture.
- a spider next to the rock.

Talk about the position of objects in the classroom.

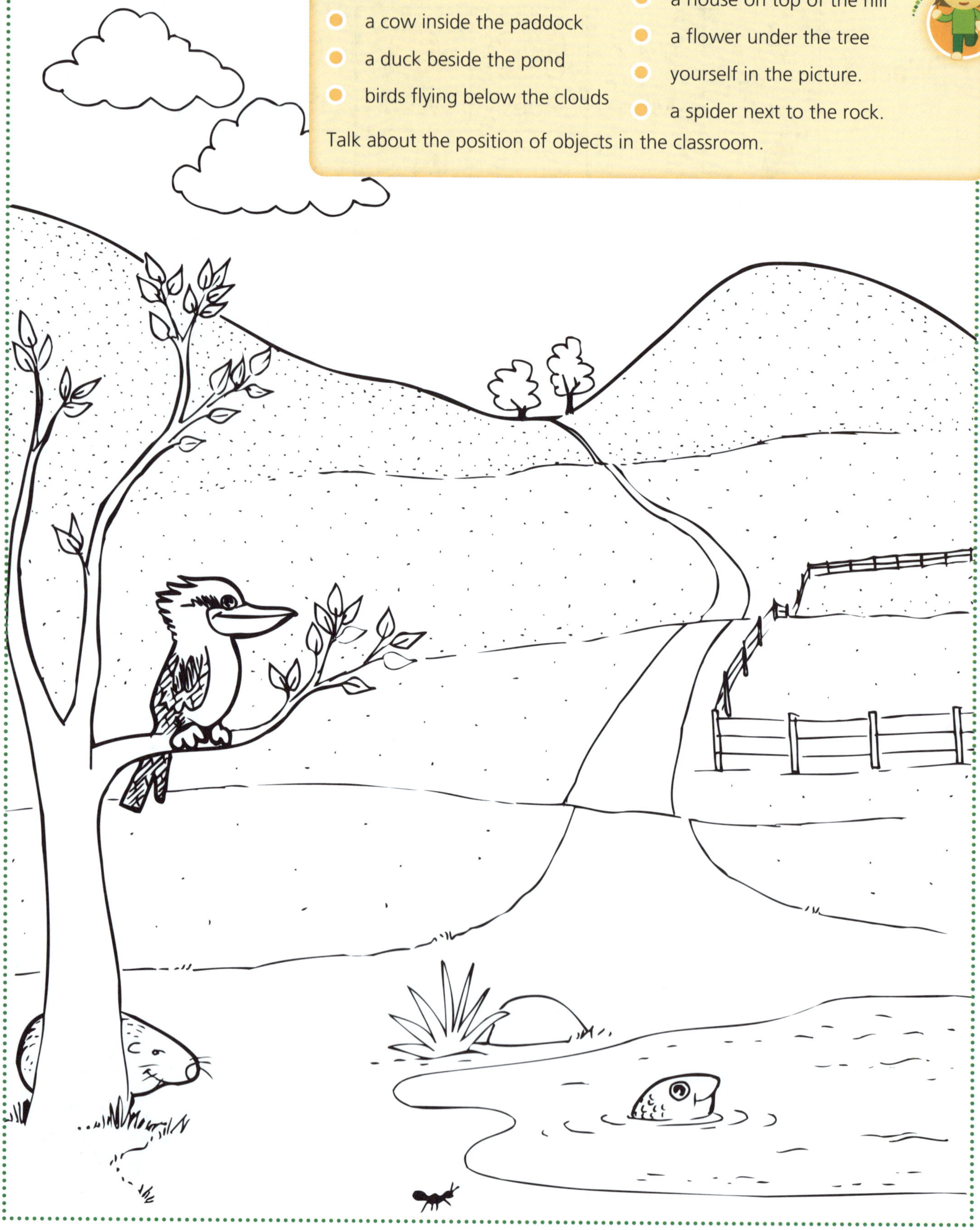

32B Pattern blocks

- Use pattern blocks to cover each picture.

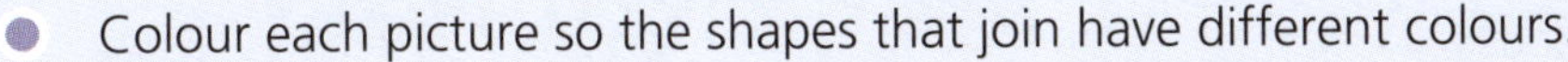

- Colour each picture so the shapes that join have different colours.

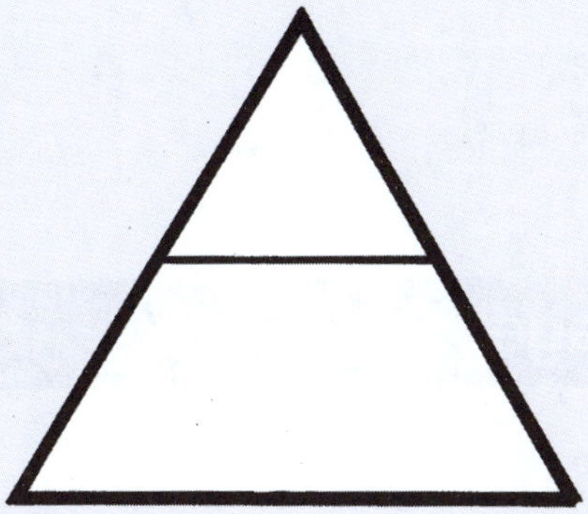

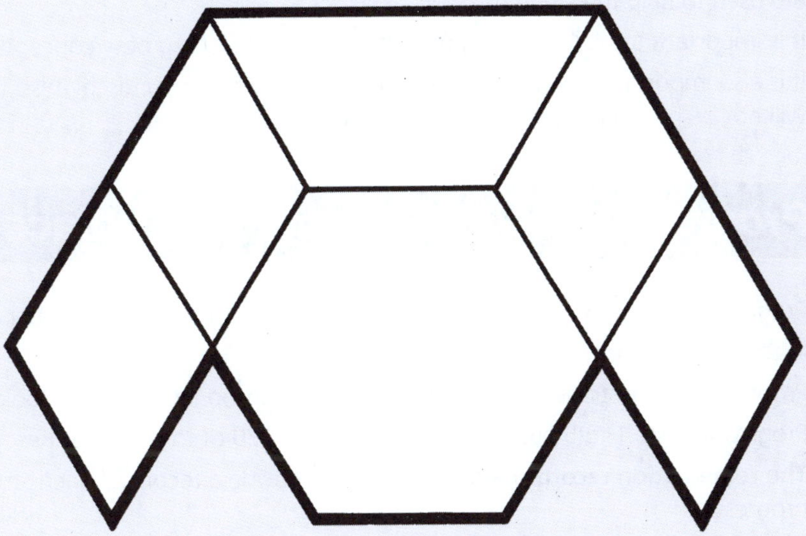

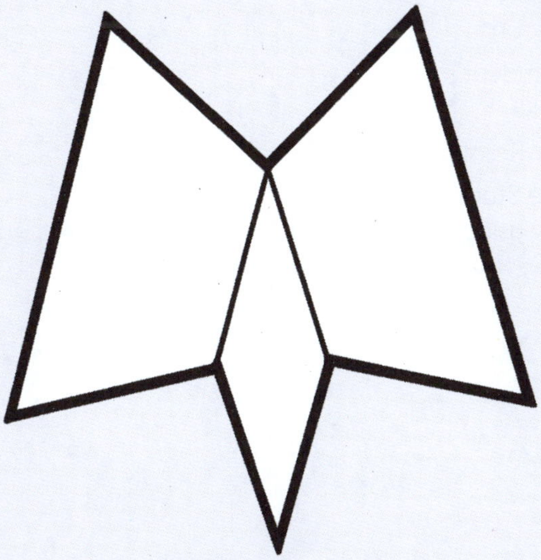

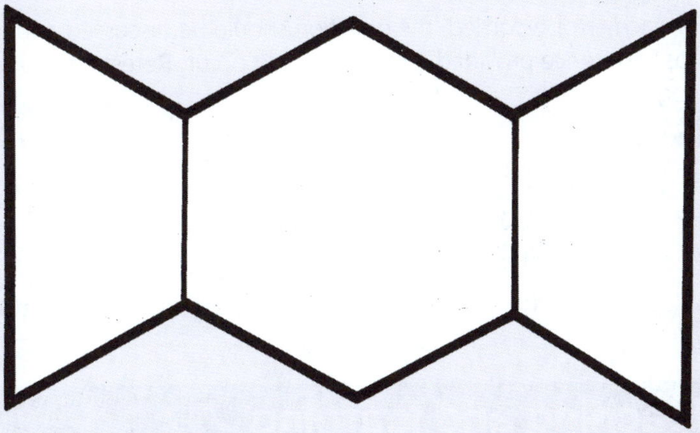

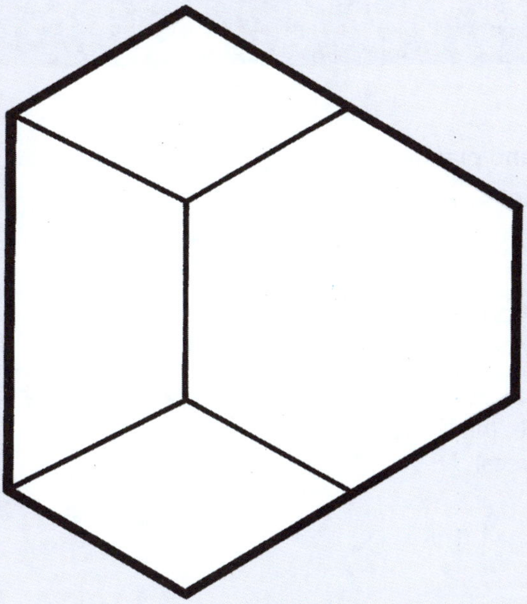

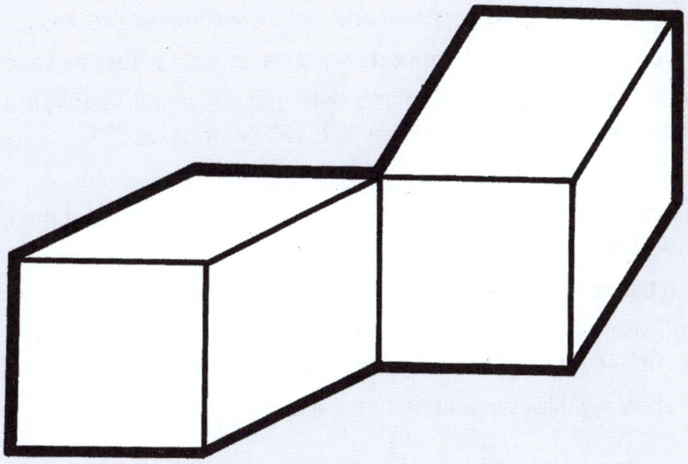

Identifying and addressing areas of need

An essential part of a teacher's role is identifying and addressing areas of student need.

This includes recognising areas where memory is fading and discovering any concepts that have been missed or misunderstood.

Testing is a great way to identify areas of need, but is only really useful when the results are used to help the student.

It is important to build a strong foundation when teaching new concepts and skills.

It is also important to revise/re-teach areas of weakness you discover so that these areas will not be barriers to the future learning of related concepts.

Progress tests and retests (see adjacent page)

Progress tests 1 to 5 are found on pages 131–154 of the online Teacher Resource.

After each test, notes and answers are supplied.

Progress test questions are cross-referenced to appropriate Student Book pages.

Progress retests 1 to 5 are found on pages 157–180 of the Teacher Resource.

The remediation records pages are used to provide a record of each student's progress.

These are found on pages 128–130 and 155–156 of the online Teacher Resource.

For each error recorded, the question should be discussed and using the Student Book cross-reference provided, practice should occur. Retesting should follow using the progress retests.

Summary

1 Test recent work.

2 Enter any mistakes in the Remediation Records.

3 Use this record to direct your revision/re-teaching.

4 Retest using the matching retest questions to ensure understanding.

Teaching and learning

Successfully teaching content and skills is a complex process.

A **good textbook** is an important tool alongside **effective teaching and planning**.

Knowledge, understanding and skills must be embedded in the student's mind so that recall continues with time. This will be done using:

(1) instruction (2) practice (3) drill (4) review.

Instruction involves explicit explanation, investigation and the use of good educational resources.

Practice forms neural pathways within the brain.

Drill strengthens neural pathways. The stronger the pathways become, the longer the understanding or knowledge is retained. 'Overlearning' prolongs recall.

Review revitalises weakened neural pathways.

Progress test 2

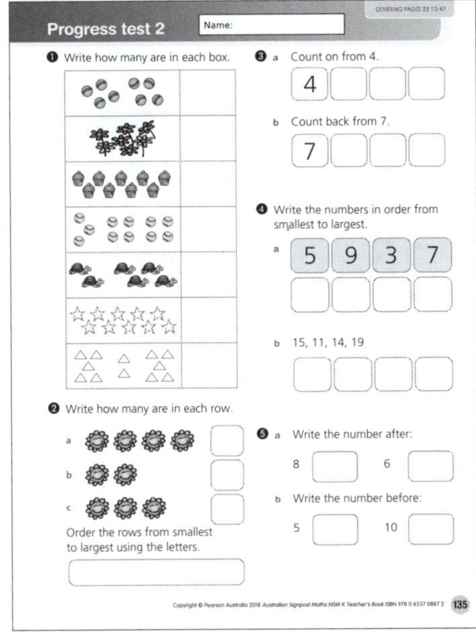
COVERING PAGES 22 TO 47

Progress test 2 Name:

1 Write how many are in each box.

2 Write how many are in each row.

a

b

c

Order the rows from smallest to largest using the letters.

3 a Count on from 4.

4

b Count back from 7.

7

4 Write the numbers in order from smallest to largest.

a 5 9 3 7

b 15, 11, 14, 19

5 a Write the number after:

8 6

b Write the number before:

5 10

Copyright © Pearson Australia 2018 *Australian Signpost Maths NSW K Teacher's Book* ISBN 978 0 6557 0887 2 135

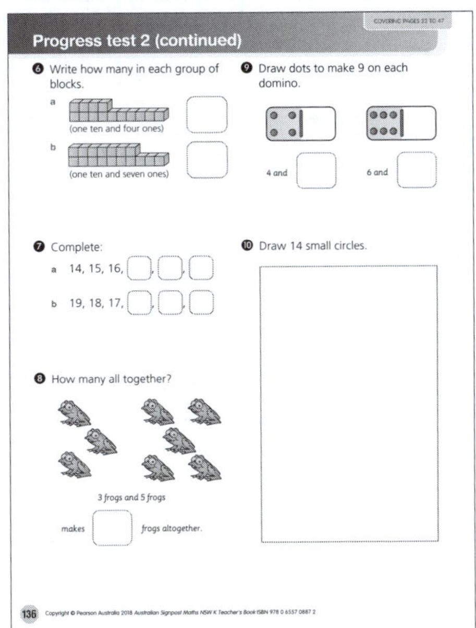
COVERING PAGES 22 TO 47

Progress test 2 (continued)

6 Write how many in each group of blocks.

a (one ten and four ones)

b (one ten and seven ones)

7 Complete:

a 14, 15, 16,

b 19, 18, 17,

8 How many all together?

3 frogs and 5 frogs

makes frogs altogether.

9 Draw dots to make 9 on each domino.

4 and 6 and

10 Draw 14 small circles.

136 Copyright © Pearson Australia 2018 *Australian Signpost Maths NSW K Teacher's Book* ISBN 978 0 6557 0887 2

Progress retest 2

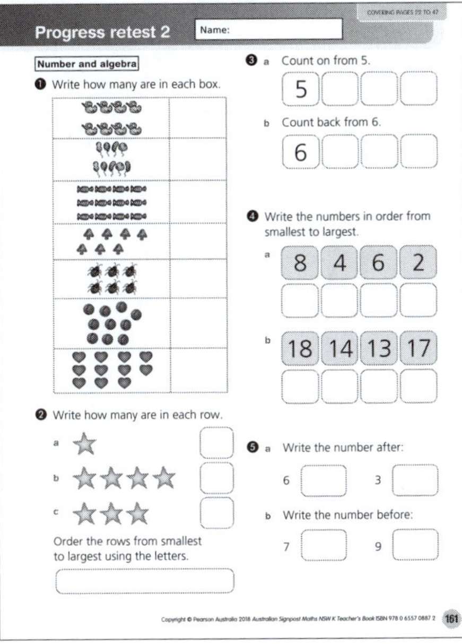
COVERING PAGES 22 TO 47

Progress retest 2 Name:

Number and algebra

1 Write how many are in each box.

2 Write how many are in each row.

a

b

c

Order the rows from smallest to largest using the letters.

3 a Count on from 5.

5

b Count back from 6.

6

4 Write the numbers in order from smallest to largest.

a 8 4 6 2

b 18 14 13 17

5 a Write the number after:

6 3

b Write the number before:

7 9

Copyright © Pearson Australia 2018 *Australian Signpost Maths NSW K Teacher's Book* ISBN 978 0 6557 0887 2 161

Notes and Answers for Progress test 2

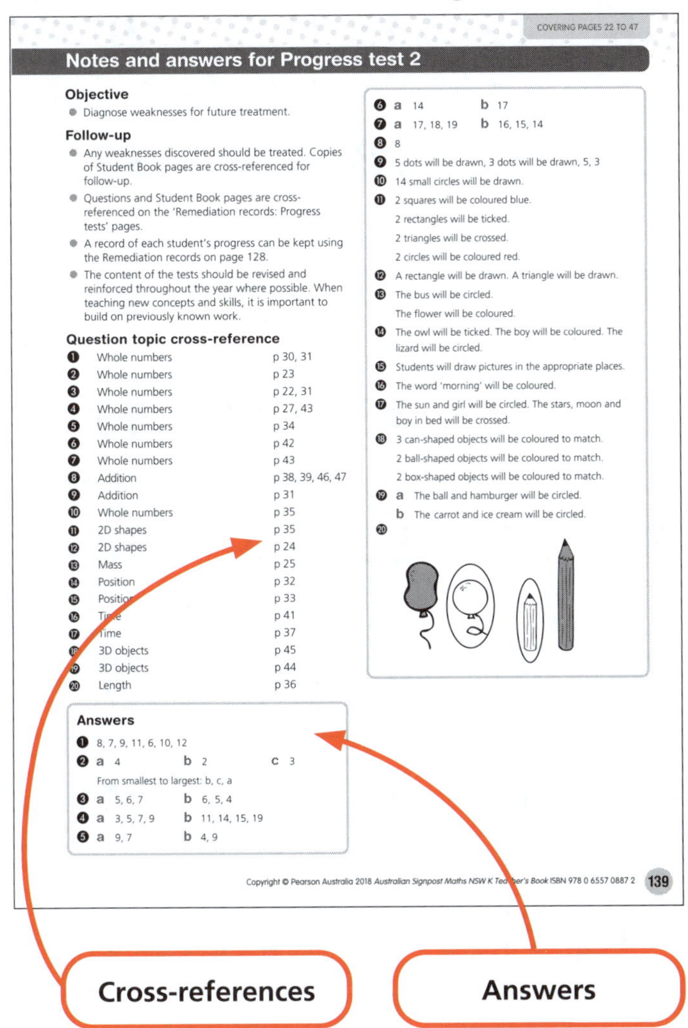
COVERING PAGES 22 TO 47

Notes and answers for Progress test 2

Objective

- Diagnose weaknesses for future treatment.

Follow-up

- Any weaknesses discovered should be treated. Copies of Student Book pages are cross-referenced for follow-up.
- Questions and Student Book pages are cross-referenced on the 'Remediation records: Progress tests' pages.
- A record of each student's progress can be kept using the Remediation records on page 128.
- The content of the tests should be revised and reinforced throughout the year where possible. When teaching new concepts and skills, it is important to build on previously known work.

Question topic cross-reference

1	Whole numbers	p 30, 31
2	Whole numbers	p 23
3	Whole numbers	p 22, 31
4	Whole numbers	p 27, 43
5	Whole numbers	p 34
6	Whole numbers	p 42
7	Whole numbers	p 43
8	Addition	p 38, 39, 46, 47
9	Addition	p 31
10	Whole numbers	p 35
11	2D shapes	p 35
12	2D shapes	p 24
13	Mass	p 25
14	Position	p 32
15	Position	p 33
16	Time	p 41
17	Time	p 37
18	3D objects	p 45
19	3D objects	p 44
20	Length	p 36

Answers

1 8, 7, 9, 11, 6, 10, 12
2 a 4 b 2 c 3
From smallest to largest: b, c, a
3 a 5, 6, 7 b 6, 5, 4
4 a 3, 5, 7, 9 b 11, 14, 15, 19
5 a 9, 7 b 4, 9
6 a 14 b 17
7 a 17, 18, 19 b 16, 15, 14
8 8
9 5 dots will be drawn, 3 dots will be drawn, 5, 3
10 14 small circles will be drawn.
11 2 squares will be coloured blue.
2 rectangles will be ticked.
2 triangles will be crossed.
2 circles will be coloured red.
12 A rectangle will be drawn. A triangle will be drawn.
13 The bus will be circled.
The flower will be coloured.
14 The owl will be ticked. The boy will be coloured. The lizard will be circled.
15 Students will draw pictures in the appropriate places.
16 The word 'morning' will be coloured.
17 The sun and girl will be circled. The stars, moon and boy in bed will be crossed.
18 3 can-shaped objects will be coloured to match.
2 ball-shaped objects will be coloured to match.
2 box-shaped objects will be coloured to match.
19 a The ball and hamburger will be circled.
b The carrot and ice cream will be circled.
20

Copyright © Pearson Australia 2018 *Australian Signpost Maths NSW K Teacher's Book* ISBN 978 0 6557 0887 2 139

Remediation records: Progress tests

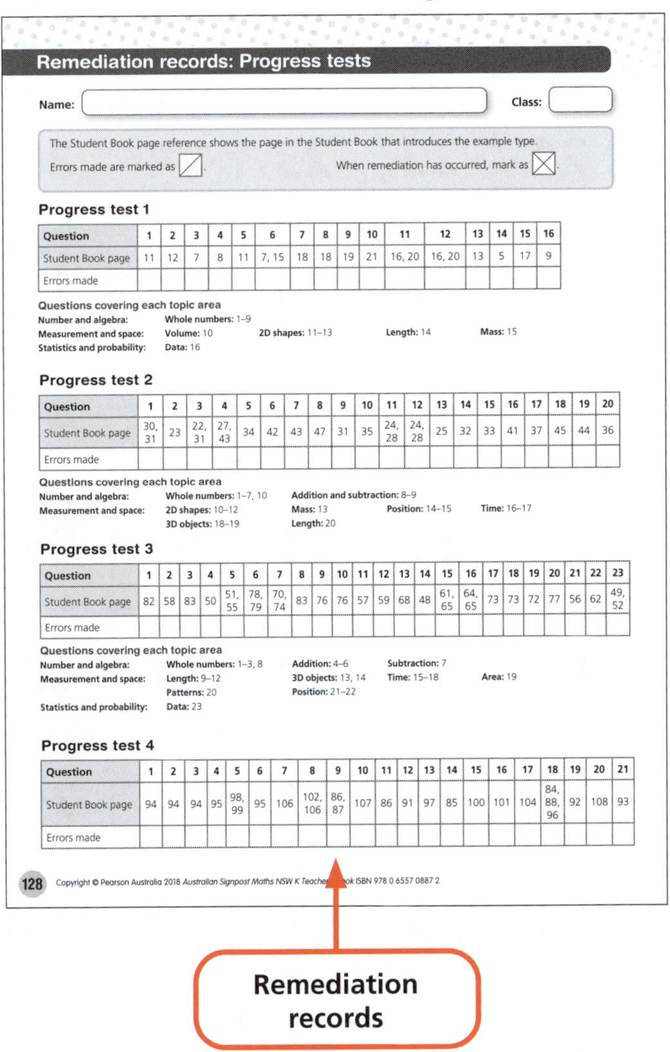
Remediation records: Progress tests

Name: Class:

The Student Book page reference shows the page in the Student Book that introduces the example type.
Errors made are marked as ◿. When remediation has occurred, mark as ☒.

Progress test 1

Question	1	2	3	4	5	6	7	8	9	10	11	12	13	14	15	16
Student Book page	11	12	7	8	11	7, 15	18	18	19	21	16, 20	16, 20	13	5	17	9
Errors made																

Questions covering each topic area
Number and algebra: **Whole numbers:** 1–9
Measurement and space: **Volume:** 10 **2D shapes:** 11–13 **Length:** 14 **Mass:** 15
Statistics and probability: **Data:** 16

Progress test 2

Question	1	2	3	4	5	6	7	8	9	10	11	12	13	14	15	16	17	18	19	20
Student Book page	30, 31	23	22, 31	27, 43	34	42	43	47	31	35	24, 28	24, 28	25	32	33	41	37	45	44	36
Errors made																				

Questions covering each topic area
Number and algebra: **Whole numbers:** 1–7, 10 **Addition and subtraction:** 8–9
Measurement and space: **2D shapes:** 10–12 **Mass:** 13 **Position:** 14–15 **Time:** 16–17
3D objects: 18–19 **Length:** 20

Progress test 3

Question	1	2	3	4	5	6	7	8	9	10	11	12	13	14	15	16	17	18	19	20	21	22	23
Student Book page	82	58	83	50	51, 55	78, 79	70, 74	83	76	76	57	59	68	48	61, 65	64, 65	73	73	72	77	56	62	49, 52
Errors made																							

Questions covering each topic area
Number and algebra: **Whole numbers:** 1–3, 8 **Addition:** 4–6 **Subtraction:** 7
Measurement and space: **Length:** 9–12 **3D objects:** 13, 14 **Time:** 15–18 **Area:** 19
Patterns: 20 **Position:** 21–22
Statistics and probability: **Data:** 23

Progress test 4

Question	1	2	3	4	5	6	7	8	9	10	11	12	13	14	15	16	17	18	19	20	21
Student Book page	94	94	94	95	98, 99	95	106	102, 106	86, 87	107	86	91	97	85	100	101	104	84, 88, 96	92	108	93
Errors made																					

128 Copyright © Pearson Australia 2018 *Australian Signpost Maths NSW K Teacher's Book* ISBN 978 0 6557 0887 2

Cross-references

Answers

Remediation records

BLM 1 Number charts

1	2	3	4	5	6	7	8	9	10
11	12	13	14	15	16	17	18	19	20
21	22	23	24	25	26	27	28	29	30

1	2	3	4	5	6	7	8	9	10
11	12	13	14	15	16	17	18	19	20
21	22	23	24	25	26	27	28	29	30
31	32	33	34	35	36	37	38	39	40
41	42	43	44	45	46	47	48	49	50
51	52	53	54	55	56	57	58	59	60
61	62	63	64	65	66	67	68	69	70
71	72	73	74	75	76	77	78	79	80
81	82	83	84	85	86	87	88	89	90
91	92	93	94	95	96	97	98	99	100

BLM 2 Ten frames

BLM 3 Place value cards

Make the number using blocks then write the number.

Put tens here.	Put ones here.

Tens	Ones

Adding two groups

and

makes

altogether

Counting on
How many fingers are hidden?

Play this game with a friend.
The other hand can be used to hide fingers.

BLM 5 Finding the difference

Use ones blocks to show each number, then write down the difference between the numbers. (A tens block can be used for 10 ones.)

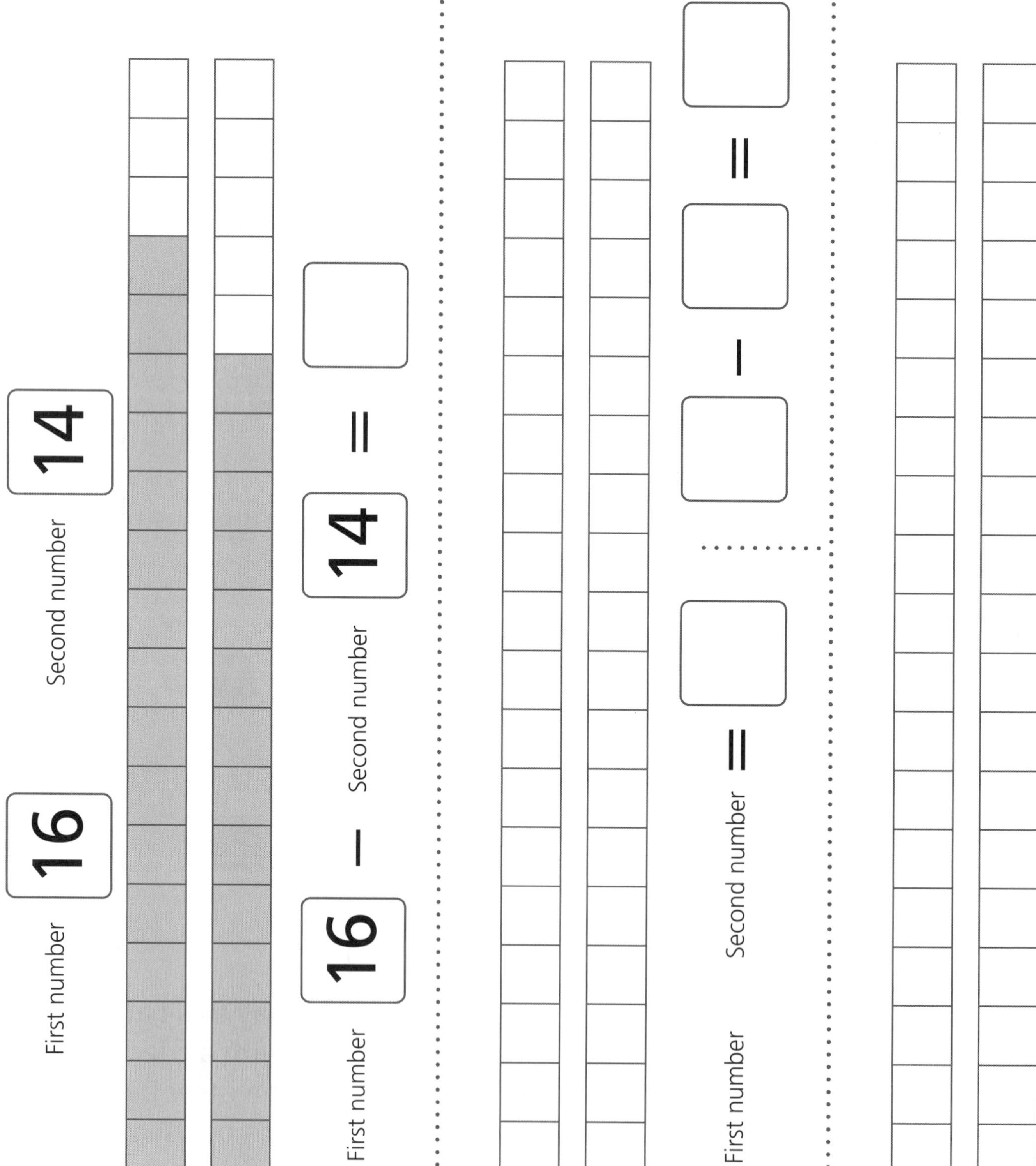

 • *AUSTRALIAN SIGNPOST MATHS NSW K* • ISBN 9780655709015

BLM 6 Sharing

BLM 7 Number bond houses

10	
1	9
2	8
3	7
4	6
5	5
6	4
7	3
8	2
9	1

9	
1	8
2	7
3	6
4	5
5	4
6	3
7	2
8	1

8	
1	7
2	6
3	5
4	4
5	3
6	2
7	1

7	
1	6
2	5
3	4
4	3
5	2
6	1

6	
1	5
2	4
3	3
4	2
5	1

5	
1	4
2	3
3	2
4	1

4	
1	3
2	2
3	1

3	
1	2
2	1

7	
	1
5	
	3
3	
	5
1	

8	
7	
	2
5	
	4
	5
2	
	7

10	
1	
	8
3	
	6
	5
6	
7	
	2
9	

BLM 8 Number bonds (addition)

Say the number bonds in a line, giving the answers as you go.
Line A would be 7 = 1 + ■, 6 = 3 + ■, 9 = 3 + ■, 8 = ■ + 5.

A

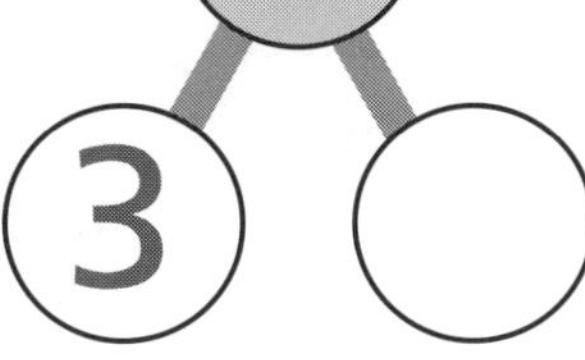

B

10
5

C

10
2

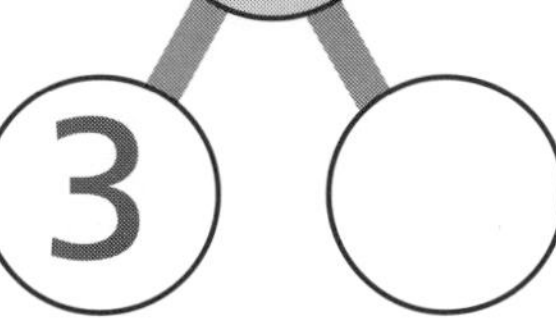

10
4

D

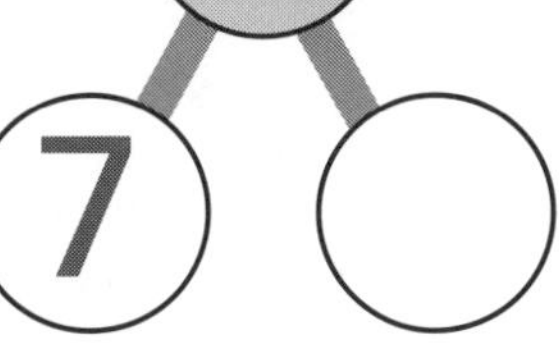

10
8

BLM 9 Using number tracks

0
1
2
3
4
5
6

Number track

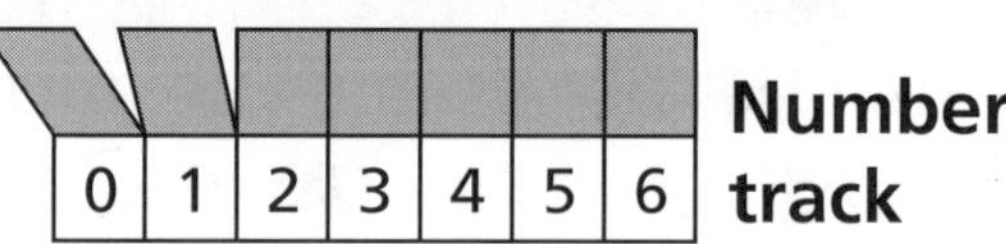

❶ Write the numbers that you will see when the flaps with arrows are lifted.

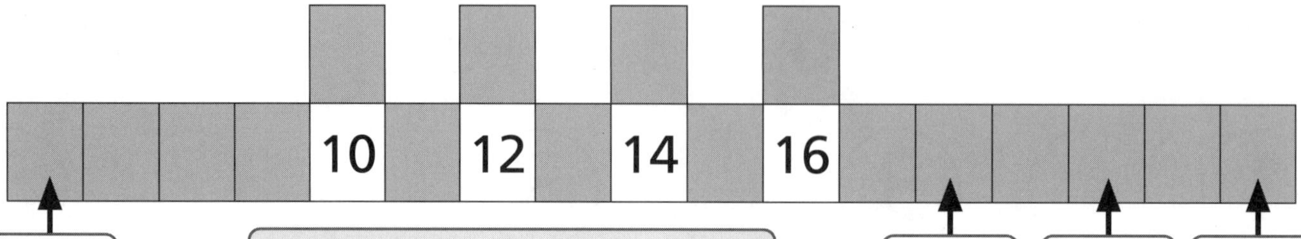

10
12
14
16

What is the difference between 12 and 14?

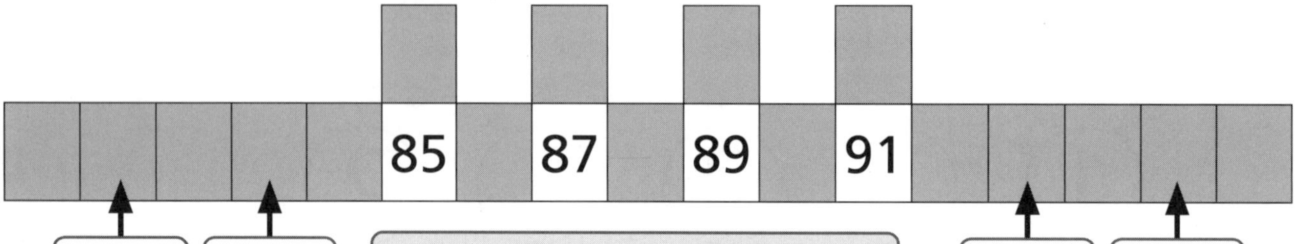

85
87
89
91

What is the difference between 87 and 89?

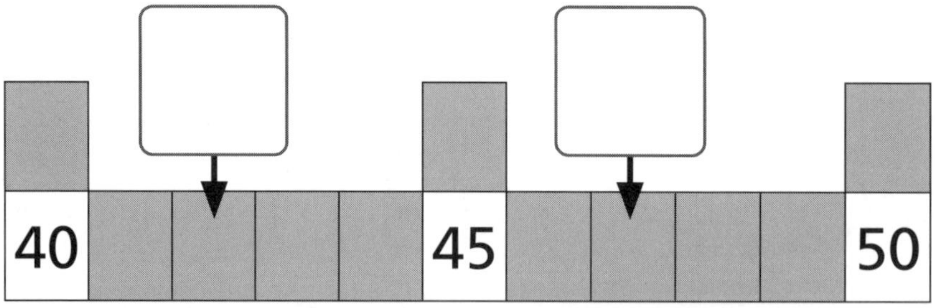

40
45
50

Is 45 halfway between 40 and 50?

Which numbers are closer to 50?

❷ Fill in the missing numbers.

40
45
50

 • *AUSTRALIAN SIGNPOST MATHS NSW K* • ISBN 9780655709015

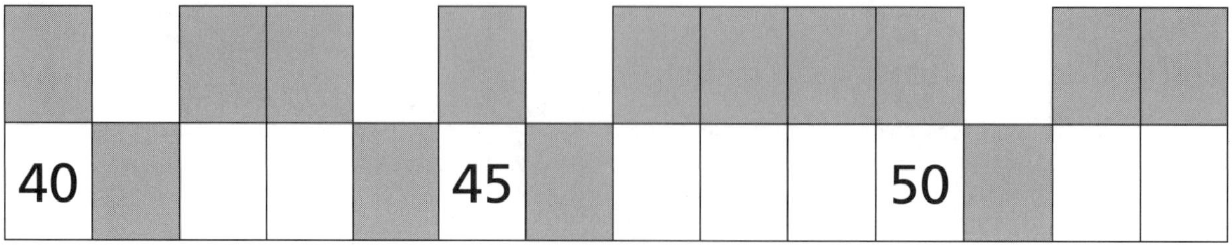